Geology of Utah's Rivers

Geology
of
Utah's Rivers

William T. Parry

THE UNIVERSITY OF UTAH PRESS
Salt Lake City

The Defiance House Man colophon is a registered trademark
of the University of Utah Press. It is based upon a four-foot-tall,
Ancient Puebloan pictograph (late PIII) near Glen Canyon, Utah.

LIBRARY OF CONGRESS CATALOGING-IN-PUBLICATION DATA

Parry, William T.
    Geology of Utah's rivers / William T. Parry.
        p.   cm.
    Includes bibliographical references.
    ISBN 978-0-87480-933-6 (pbk. : alk. paper)   1. Geology—Utah.
2. Rivers—Utah.   I. Title.
    QE169.P37 2008
    557.92—dc22   2008010528

Contents

CONTENTS

Maps

Acknowledgments

Most of my professional career was spent teaching in the classroom or doing scientific research as a geochemist in the laboratory. Near the end of my teaching career, I was assigned by the department chair to teach a course titled "Geology and Scenery of Utah." This was a course that had previously been taught by many of the great geologists of Utah including Armand Eardley, William Lee Stokes, Frank DeCourten, and Genevieve Atwood, and I didn't think I was up to the task. However, I taught the course and associated field trips for the next ten years until my retirement in 2001, and through it became fascinated with the story of Utah's rivers.

I then began fly fishing under the tutelage and encouragement of L. H. (Chuck) Wullstein, who suggested that fishing enthusiasts might like to read about the geology of the rivers they fish. The combination of his suggestion and encouragement and my fascination led to the present book.

My thanks to each one of those great geologists who preceded me and to all of my teachers and students from whom I learned much. I am grateful to Chuck Wullstein for the gift of fly fishing and for his encouragement and patience. The careful review and detailed editing of Dr. Kathleen Nicoll improved the presentation greatly. I acknowledge with gratitude the comments and suggestions of William F. Case and the patience and critical editing of Gayle Parry.

Dr. Ron Blakey, Department of Geology, Northern Arizona University, generously permitted use of paleogeographic maps 2 through 7. The shaded relief maps were prepared with digital elevation data from the U.S. Geological Survey using the freeware computer program MICRODEM produced by Professor Peter Guth, Oceanography Department, U.S. Naval Academy.

Geology of Utah's Rivers

A river, though, has so many things to say
that it is hard to know what it says to each of us.

— NORMAN MACLEAN, *A River Runs Through It*

.

Introduction

A substantial portion of the geologic and cultural history of Utah involves rivers, and the effects of running water are evident in the present landscape. This book is about the geology of the rivers that have shaped the present landscape, the ancient river deposits of sand and gravel that are now part of the mountains, and the mythical rivers that inspired early exploration. The ancient rivers are gone now and so are the mountains that gave rise to them. Yet the sedimentary rocks that formed from the materials transported in the paleorivers remain. For example, a huge river transported material from mountains in Colorado and Texas westward to the ocean in western Utah and Nevada during Triassic time. The sedimentary rocks that formed from deposits of that river are now known in Utah as the Chinle Formation, which also contains the remains of ancient fish that inhabited its waters, the first dinosaurs, and ancient forests that lined the riverbanks.

Many national parks and monuments in Utah are located along rivers. Zion National Park is on the Virgin River; Bryce Canyon National Park is on the headwaters of the Paria River; Capitol Reef National Park is on the Fremont River; and Canyonlands National Park is on both the Green and Colorado rivers. You'll find Dinosaur National Monument on the Green River, Rainbow Bridge National Monument on the Colorado, and Grand Staircase–Escalante National Monument on the Escalante and Paria rivers.

Highways in Utah follow river corridors. Interstate 84 and Interstate 80 follow the Weber River; U.S. Highway 89 follows the Sevier River and the East Fork of the Virgin River; U.S. 40 follows the Strawberry River; and U.S. 189 and Utah 150 follow the Provo River.

This book is not only about the ancient rivers; it is also about rivers that exist today. Many of today's rivers have an ancient past that can be deduced by careful examination of the landforms, the deposits, and the course of the rivers. Ancient and modern rivers preserve a record of climate change recorded in the capacity to transport sediment and in deepening of their valleys. The record of valley deepening is preserved in river terraces that are often exposed above the present river level. New age determination techniques such as cosmogenic dating have recently been developed to measure the age of these terraces and other surfaces and thus determine the evolution of a river valley. Studies of modern and ancient river systems may also provide a record of climate change (Milhous 2005).

Utah is the second driest state in the United States, surpassed only by Nevada. No rivers flow through Nevada today. Nevada is internally drained. Its rivers have no outlet to the sea, but instead terminate in lakes. The Colorado River defines Nevada's southern boundary with Arizona, and the Virgin River that originates in Utah joins the Colorado at Lake Mead in southern Nevada. Otherwise, most of the rivers that originate in Nevada flow into and terminate in basins within the state. The Truckee River begins in the Sierra Nevada Mountains and ends in Pyramid Lake; the Carson River ends in Carson Lake; the Humboldt and Reese river end in Humboldt Lake; the Muddy and Virgin rivers end in Lake Mead; the Franklin River ends in Ruby Valley; the White River ends in Pahranagat Valley; and the Walker River ends in Walker Lake. The Owyhee and Jarbidge rivers, the only exceptions, flow out of Nevada into Idaho.

The high mountains in Colorado and Wyoming give rise to a very large river system in Utah. The Green and Colorado rivers and their tributaries flow through the arid Colorado Plateau region of Utah. Most of southwestern Wyoming is drained by the Green River, which flows into the Colorado and eventually into the Pacific Ocean. The high Wind River Mountains of Wyoming provide the start for the Green River that traverses the Uinta Mountains before joining the Colorado River in Utah. Major Colorado River tributaries, including the San Juan, the White, the Yampa, and the Dolores rivers, originate in the high mountains of Colorado. The Provo, Bear, Weber, and Duchesne rivers originate in the Uinta Mountains, and the Sevier River begins on the high, western margin of the Colorado Plateau.

Utah has some remarkable rivers. The Bear River is the longest river in the United States that does not empty into an ocean. The Colorado River crosses the most structural barriers in the Western Hemisphere and crosses the most physiographic provinces. The Sevier River is the longest river entirely within the state. The rivers in Utah exhibit evidence of drainage reversals, river piracy, superposition, antecedence, and entrenched meanders—features which have been produced during the history of each river system.

Described in this book are the mountain-building events that provided a source for the rivers, the physiographic provinces of Utah that give the modern rivers their character, and the deposits of sand and gravel that formed as ancient rivers flowed from ancient mountains during Utah's geologic history. The course of the modern rivers and the rock strata over which they flow are described together with the processes of consequence, antecedence, and superposition that produced their course. The description of rivers is completed with the historical account of rivers that did not exist, the mythical rivers of Utah. Fossil and modern fish fauna are described to provide a unique guide for discussion of some important aspects of rivers, such as a former connection of the Green River with the North Platte River that is also indicated by geomorphic evidence.

This book is meant as a guide to the geology of the rivers for hikers, fishermen, boaters, and others. It is not a river guide suitable for bend-to-bend, rapid-by-rapid white-water boating, nor does it locate the best fishing holes. More detailed river and fishing guides should be consulted for this purpose. River guides include Baars (1987), Baars and Stevenson (1986), Belknap and Belknap (1974), Belknap and Evans (1995), Evans and Belknap (1973, 1974), Hansen (1996), Hayes (1969), Hayes and Santos (1969), Hayes and Simmons (1973), Mutschler (1969a, 1969b), and Nichols (2002) (a guide to every river in Utah with a possibility of boating). Fishing guides for Utah rivers include Cook (2000), Demoux (2001), and Prettyman (2001).

Shaded relief maps and maps of the geology along each stretch of river accompany the descriptions of present-day rivers in Utah, beginning with the Green River in Chapter 6. The geologic maps are modified from Stokes et al. (1961–1963). The maps are modified by updated stratigraphic names from Hintze (1988) and by grouping formations exposed on steep cliffs forming narrow outcrop patterns. An index map to the locations of individual riverside maps is provided on Map 1. Each individual map is accompanied by an index showing the location of the stretch of river that is depicted. Figure 1 shows the geologic time scale and some of the major events that shaped the rivers.

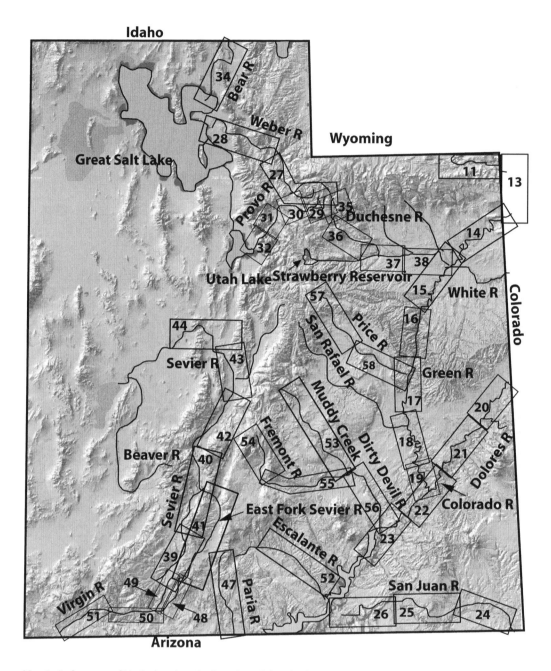

Map 1. Index map of Utah showing the location of the shaded relief and geologic maps of Utah rivers discussed in the text.

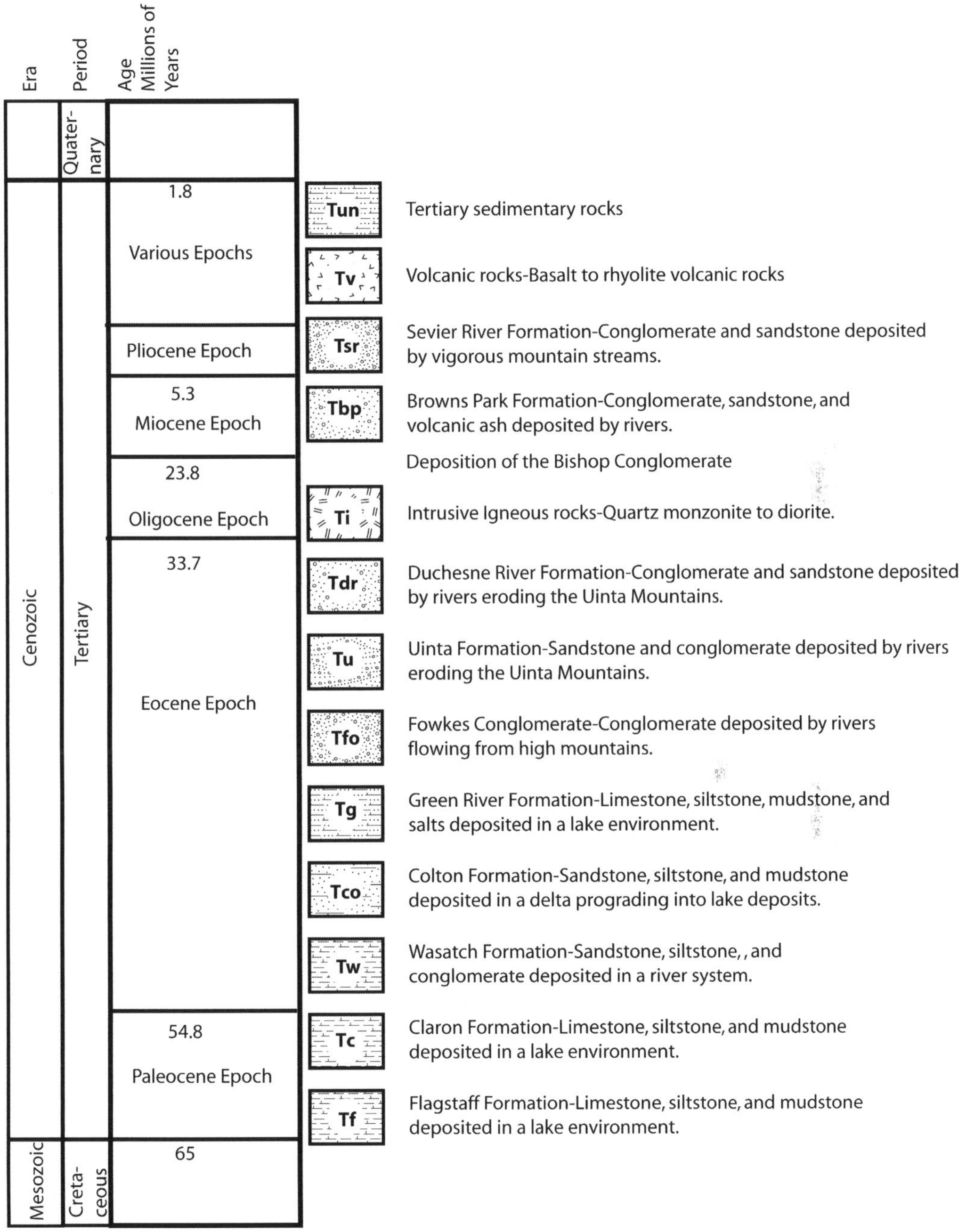

Era	Period	Age Millions of Years		

The chart content:

Era: Cenozoic (spanning), Mesozoic (bottom)

Period: Quaternary, Tertiary, Cretaceous

Age Millions of Years / Epochs:

1.8

Various Epochs

Pliocene Epoch

5.3
Miocene Epoch

23.8
Oligocene Epoch

33.7

Eocene Epoch

54.8

Paleocene Epoch

65

Rock units and descriptions:

Tun — Tertiary sedimentary rocks

Tv — Volcanic rocks-Basalt to rhyolite volcanic rocks

Tsr — Sevier River Formation-Conglomerate and sandstone deposited by vigorous mountain streams.

Tbp — Browns Park Formation-Conglomerate, sandstone, and volcanic ash deposited by rivers.

Deposition of the Bishop Conglomerate

Ti — Intrusive Igneous rocks-Quartz monzonite to diorite.

Tdr — Duchesne River Formation-Conglomerate and sandstone deposited by rivers eroding the Uinta Mountains.

Tu — Uinta Formation-Sandstone and conglomerate deposited by rivers eroding the Uinta Mountains.

Tfo — Fowkes Conglomerate-Conglomerate deposited by rivers flowing from high mountains.

Tg — Green River Formation-Limestone, siltstone, mudstone, and salts deposited in a lake environment.

Tco — Colton Formation-Sandstone, siltstone, and mudstone deposited in a delta prograding into lake deposits.

Tw — Wasatch Formation-Sandstone, siltstone, , and conglomerate deposited in a river system.

Tc — Claron Formation-Limestone, siltstone, and mudstone deposited in a lake environment.

Tf — Flagstaff Formation-Limestone, siltstone, and mudstone deposited in a lake environment.

Figure 1. The sedimentary and igneous rocks exposed along Utah's rivers. The list includes the geologic and absolute age of each interval of time. Each rock unit is shown with the map pattern used to indicate outcrop areas on the geologic maps of each river section.
1. Cenozoic Era.

Era	Period	Age Millions of Years
Cenozoic	Tertiary	
		65
Mesozoic	Cretaceous	Various Ages
		Maastrichtian Age
		71.3 Campanian Age
		83.5 Santonian Age
		85.8 Coniacian Age
		89 Turonian Age
		93.5 Cenomanian Age
		99 Albian Age
		112 Aptian Age
		121 Neocomian Age
	Jurassic	144

K — Cretaceous sedimentary rocks undifferentiated.

TKnh — North Horn Formation-Conglomerate, sandstone, and siltstone deposited in a river system.

Kpr — Price River Formation- Conglomerate, sandstone, and siltstone deposited in a river system.

Kwa — Wahweap Sandstone

Khfe — Frontier, Henefer, and Echo Canyon conglomerate-Conglomerate deposited in vigorous mountain streams flowing from Cretaceous mountains.

Kbh — Blackhawk Formation-Coastal plain sandstone, siltstone, shale, and coal.

Km — Mancos Shale-Marine shale deposited in the interior sea that connected the Gulf of Alaska with the Gulf of Mexico.

Kdcm — Dakota Sandstone and Cedar Mountain Formation-Conglomerate and sandstone deposited in river and floodplain systems.

Kd — Dakota Sandstone-Sandstone and siltstone deposited in a river system.

Kdbc — Dakota Sandstone and Burro Canyon Formation-Sandstone, siltstone, and mudstone deposited in a river system.

Sevier mountain building begins

Figure 1 continued. Cretaceous Period.

Era	Period	Age Millions of Years		Description
Mesozoic	Cretaceous			
	Jurassic	144	J	Jurassic formations undivided
			Jm	Morrison Formation-Sandstone, siltstone, mudstone, and conglomerate accumulated in river and lake systems.
			Jscu	Summerville and Curtis Formations
			Jsu	Summerville Formation-Siltstone, mudstone, and gypsum deposited in a near-shore marine environment.
			Jc	Curtis Formation-Sandstone and mudstone deposited in a near-shore marine environment.
			Ja	Arapien Shale-Marine shale, siltstone, gypsum, anhydrite, and salt. Equivalent to the Twin Creek and Carmel.
			Jtc	Twin Creek Limestone-Marine limestone. Equivalent to the Arapien and Carmel.
			Jgc	Glen Canyon Group-Wind deposited sand. Equivalent to the Navajo, Kayenta, and Wingate.
			Jca	Carmel Formation- Marine limestone, shale, siltstone, gypsum, and anhydrite. Approximately equivalent to the Arapien and the Twin Creek.
			Je	Entrada Formation
			Jna	Navajo Sandstone-Wind deposited sand. Equivalent to the Nugget.
			Jn	Nugget Sandstone-Wind deposited sand. Equivalent to the Navajo.
	Triassic	206		

Figure 1 continued. Jurassic Period.

Era	Period	Age Millions of Years
Mesozoic	Jurassic
Mesozoic	Triassic	206
Mesozoic	Triassic	248
	Permian

Ṟ -J — Triassic and Jurassic formations undivided

Ṟ — Triassic formations undivided

Ṟkw — Kayenta and Wingate formations

Ṟk — Kayenta Formation-Continental shale, siltstone, and sandstone accumulated in a river system.

Ṟkcm — Kayenta, Moenave, and Chinle Formations

Ṟc-w — Chinle to Wingate formations-Wingate is a wind deposit of sand that forms steep cliffs with narrow outcrop pattern.

Ṟc — Chinle Formation-Continental shale, siltstone, and sandstone accumulated in a river system. Approximately equivalent to the Ankareh.

Ṟc-m — Chinle and Moenkopi formations

Ṟa — Ankareh Formation-Continental shale, siltstone, and sandstone accumulated in a river system. Approximately equivalent to the upper Moenkopi and Chinle.

Ṟt — Thaynes Limestone- Marine limestone accumulated on the continental margin. Approximately equivalent to the middle Moenkopi

Ṟw — Woodside Shale-Marine red shale accumulated on the continental margin. Approximately equivalent to the lower Moenkopi.

Ṟm — Moenkopi Formation-Marine red shale and limestone accumulated on the continental margin. Approximately equivalent to the Woodside Shale

Figure 1 continued. Triassic Period.

Era	Period	Age Millions of Years
Mesozoic	Triassic	
Paleozoic	Permian	248
		290
	Pennsylvanian	
		323
	Mississippian	
		354

Ppc Park City Formation-Limestone and phosphatic shale

Pk Kaibab Formation-Limestone and dolomite.

Pcu Cutler Group-White Rim Sandstone, Organ Rock Shale, Cedar Mesa Sandstone, and Elephant Canyon Formation.

Pcm Cedar Mesa Sandstone-Member of the Cutler Group-Wind deposited sandstone.

IPPr Elephant Canyon Formation-Member of the Cutler Group

IPw Weber Sandstone-Shoreline sandstone and wind deposited sandstone

IPm Morgan Formation-Limestone, siltstone, and sandstone.

IPmr Morgan and Round Valley Limestone

IPo Oquirrh Formation-Sandstone, limestone, siltstone, and shale deposited in the Oquirrh Basin.

IPh Honaker Trail Formation-Sandstone, siltstone, limestone.

IPp Paradox Formation-Sandstone, siltstone, shale, limestone, anhydrite, and salt.

Mun Mississippian undivided

Mb Brazer Dolomite-

Mm Madison Limestone-Limestone and dolomite deposited on the continental shelf.

Figure 1 continued. Mississippian, Pennsylvanian, and Permian Periods.

Era	Period	Age Millions of Years
	Devonian	354
	Ordovician-Silurian	417
		490
Paleozoic	Cambrian	
		543
Precambrian	Proterozoic	
		2500
	Archean	
		4550

Pal — Paleozoic formations

Du — Devonian formations

OS — Ordovician-Silurian formations

Є — Cambrian formations

Єo — Ophir Formation-Shale and a middle limestone deposited on a passive continental margin

Єt — Tintic Quartzite-Quartzite formed from sand deposited in a shoreline environment

Єl — Lodore Formation-Sandstone and shale deposited in a near shore marine environment

PЄ — Big Cottonwood Formation-Quartzite and argillite formed from sediments deposited in a tidal estuary.

PЄu — Uinta Mountain Group-Sandstone, shale, and conglomerate deposited by rivers in a rift valley.

PЄrc — Red Creek Quartzite-Quartzite, schist, and amphibolite formed from metamorphism of sandstone, limestone, and shale deposited in a shallow marine environment offshore of the Archean continent.

PЄ — Farmington Canyon Complex-Schist and gneiss intruded by pegmatite.

Figure 1 continued. Precambrian Era, Cambrian, Ordovician, Silurian, and Devonian Periods.

How Rivers Function

Much of our knowledge of rivers goes back to the earliest days of the study of geology. River systematics were formalized in Sir Charles Lyell's book *Elements of Geology*, which was published in eleven editions between 1830 and 1872. The first edition in three volumes was published in 1830 to 1833. In those years, studies of individual rivers were less scientific and less quantitative than today's earth sciences research.

Erosion and deposition by running water are the most important agents that shape the Earth's landscape. Rain falls on the land surface and collects in channels, often on steep mountain slopes. Channel width, depth, and the ratio of width to depth are parameters that increase systematically with increasing drainage area and volume of water that must be accommodated in a river channel. For example, the gradient of steep mountain rivers may be greater than 250 feet per mile, but the gradient decreases downriver to 0.5 foot per mile or less near the mouth.

Gradient, cross-section of the channel, velocity of the flowing water, volume of water that must be accommodated, and the quantity of sediment that is transported affect river flow. These characteristics of a river change systematically along its course. The discharge increases, the cross-sectional area of the channel increases, and the velocity increases as the gradient decreases downriver. The number of tributaries also decreases as their length increases. The increase in velocity as gradient decreases may be surprising, but more water is added with each tributary entering the river, so more water must flow through the river channel, adding an increment to the velocity. Steep and vigorous mountain rivers that seem to flow with great velocity are deceptive. The water must negotiate many obstacles to flow including all of the irregularities of the channel, boulders in the channel, and debris, all of which add friction to the running water and lower the velocity. As gradient decreases and rivers deposit sediment in their beds, the channel may begin to meander. These meanders usually migrate downriver. If a river is unable to move its entire available load, a braided pattern may develop in which water divides and reunites with two or more adjacent but interconnected and constantly shifting channels.

Water flow velocity controls the riverbed shear stress and is the dominant control on the maximum particle size that a river can transport. The relationship between particle size and water velocity is not linear throughout the full range of velocities. For example, a velocity of about 7 feet per second is required to move cobbles (particles that are 4 inches in diameter). A velocity of 2 feet per second will move coarse sand (particles that are at least 0.04 inches in diameter), and a velocity of 0.03 feet per second will move only fine sand, silt, and clay (particles that are smaller than 0.04 inches in diameter). The velocity of running water in a river controls erosion, transport, and deposition of various particle sizes.

Rivers will deposit the sediment that they carry in response to decreased gradient, reduced discharge as in the waning flood stages, or increased sediment load. Deposits are often left in broad flood plains following flood stages. The renewed ability of a river to erode resulting from an increase in velocity, an increase in discharge, or a decrease in sediment load may enable a river to erode into its own flood plain leaving river terraces as remnants of the abandoned flood plain.

River erosion and the landforms that are produced are responsible for most of the landscapes we view today. The evolution of a landscape is reflected in topography, hydrology, and sediment transport linked to the underlying geological setting. Rivers and river channels have an intimate relationship with the geology of the surfaces over which they flow. The slope of a newly uplifted or emergent land surface determines the initial course of a consequent river.

A subsequent river preferentially erodes weak rock strata and its course therefore is determined by the patterns of exposure of the weak, easily eroded rocks. The course of the river is often a reflection of underlying rock structure such as folds or faults that expose belts of weak, soft, easily eroded rock. Tilted rocks on the flanks of a fold cause a subsequent river to follow the bands of more easily eroded rock.

An antecedent river is older than the structure that it crosses. It maintains its course across the rocks that have been raised across its course, for example, a steep narrow gorge is formed across a structural ridge raised in a river's path. A superposed river cuts downward through overlying rocks, placing its channel on underlying rocks with different lithology or structure. A superposed river course is gradually lowered down to underlying structure, while the structure is gradually raised in the course of an antecedent river. The course of the river is not determined by the structure of the rocks over which a superposed or antecedent river is flowing. With time and erosive power, a superposed river course can change to reflect the underlying structure.

Distinctive river patterns often develop in certain geological circumstances. The pattern of river drainage is as diverse as the geology of underlying rocks. For example, a dendritic pattern develops on massive rocks of uniform resistance to erosion or on flat-lying sedimentary strata; a parallel drainage pattern develops on sloping surfaces of homogenous rock; a radial pattern forms on a topographic high such as a dome or volcano where rivers flow away from the highest elevation. A rectangular drainage pattern with near right-angle bends forms because the drainage follows joints, faults, and fractures. A trellised pattern forms when long, resistant ridges of folded rocks force rivers to follow the easily eroded rocks between the ridges, but at times the rivers cross the ridges. Annular rivers form segments of circles flowing around some geological obstacle, and centripetal drainages converge toward a center.

River systems are ordered in several ways. The number of tributaries decreases systematically downriver, length of tributaries increases downriver, the gradient decreases exponentially downriver, channels widen and deepen downriver, and the size of a valley carved by a river is proportional to the size of the river. Utah has a number of examples of underfit rivers, rivers that are too small for the size of the river-cut valley in which they flow. Interpretation of events leading to underfit rivers is an important part of river history.

Rivers erode their valleys and deposit the resulting sediment in response to changes in discharge, load, gradient, and base level. Base level is the lowest elevation at which a stream can erode its bed. Sea level is usually base level, but lakes may form a temporary base level. River terraces are formed when a river cuts downward through its own deposits in response to increased velocity due to lowering of base level or a decreased sediment load. The history of river terraces can provide important clues to the history of down-cutting by a river. Alluvial fans (Figure 2) result when mountain rivers reach the mountain front and channel width changes, decreasing velocity and causing deposition of sediment.

Gradients in the headwaters of a river are usually high, yielding high velocities and high turbulence, and enhancing the capacity for erosion. Rivers thus tend to lengthen their valleys headward, called headward erosion. A river that is vigorously lengthening may capture other drainage channels.

Geomorphologists, those geologists who study landforms, have developed a complex terminology for rivers. Youthful rivers flow from high topography to lower topography. The initial development of river channels may be controlled largely by chance. A dentric drainage pattern

Figure 2. Modern alluvial fans accumulating at the base of the Kun Lun Mountains near Pishan in northwest China, about 174 miles southeast of Kashgar along the ancient Silk Route in the Taklamakan desert of the Xinjian Province. This image was acquired by the Spaceborne Imaging Radar and is provided by NASA.

would develop where the river and its tributaries resemble the pattern of branching in the veins of a leaf or branches of a tree. Such rivers are called consequent rivers because they flow downslope as a consequence of the topography. However, with time, erosion into underlying bedrock may produce some significant changes in the course of a river. Bedrock may include inclined layers of sediment that differ in resistance to erosion. River channels will tend to follow the more easily eroded beds. The resulting pattern of river channels is similar to the pattern of latticework constructed to support growing vines such as a rose trellis. Obsequent rivers flow in a direction opposite to the dip of beds (Sparks 1986). The terminology of rivers gets even more complicated. Resequent rivers are evolved consequent rivers. For example, a consequent rivers forms on a newly formed anticline, then erodes into the anticline and becomes resequent or obsequent when flow is opposite to the dip of beds on the flanks of the anticline.

Consequent rivers follow initial irregularities in the land surface after a new uplift begins an early stage of the erosion cycle. The initial slope of the land determines the course of a consequent river. Over time, rivers tend to follow more easily eroded rocks. Subsequent rivers follow courses through some weakness such as weak rock layers or faults. Subsequent rivers are developed by headward erosion under the guidance of weak structure. Insequent rivers do not depend on initial slope or depression or soft rocks; instead, they develop on homogenous rock and their course develops by chance.

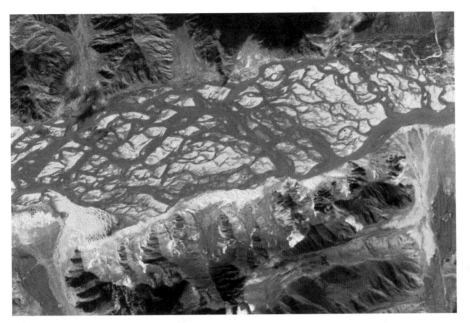

Figure 3. A braided river. A 10-mile stretch of the Brahmaputra River about 35 km south of Lhasa is intricately braided from heavy sediment load in an east-west valley between the Tibetan Plateau to the north and the Himalaya Mountains on the south. The Earth Sciences and Image Analysis Laboratory at Johnson Space Center provided this image taken from the International Space Station.

The course of rivers is determined by a complex interplay between antecedence, superposition, headward erosion, and river piracy. John Wesley Powell in 1875 distinguished three kinds of valleys on the basis of their geologic history. Consequent valleys have courses directly inherited from a bedrock surface formed by folding or tilting. The axis of a newly formed syncline becomes a river course. The axis of an anticline becomes a drainage divide. Radial drainage developed from a volcanic pile and uplift in the San Juan Mountains is consequent drainage. Antecedent valleys persist on land surfaces where folds or other displacements form after a consequent drainage has formed. The antecedent river deepens its channel and keeps pace with the uplift. Superimposed valleys form where younger sediments on which a consequent drainage forms bury previously folded or tilted rocks. The superimposed drainage cuts downward into buried structures.

Gilbert (1877) gave a clear description of river capture and superposition. Drainage may become superimposed from marine deposits that bury old structures, from alluvial or other terrestrial deposits that bury old structures, or from erosion surfaces truncating older structures. Most river capture occurs when a river with a steeper gradient lengthens its valley by headward erosion and intercepts a second river with a lower gradient.

Because of mountain and plateau uplift, Utah landscapes present many examples of river capture.

Two additional characteristics of rivers will be useful in the following descriptions. Meandering rivers generally develop on gentle slopes in wide valleys. The meanders migrate across the valley eroding the floodplain deposits on the cut bank (outside bend) and building floodplain on the point bar side (inside bend). The meanders help dissipate the hydraulic potential energy of the running water. Braided rivers form when the water cannot transport the sediment load. As the velocity decreases, sediment is deposited on the channel floor creating extensive bars that separate the channel into a series of smaller channels that constantly shift in position (Figure 3). Braided rivers usually have a higher gradient than meandering rivers and are shallower in depth than meandering rivers. Braided rivers have wide, shallow channels, weak banks that are easily eroded to widen the channel, and reduced vegetation.

A barbed tributary is a tributary that enters another river in an upriver direction. The barbed tributary thus flows "against the grain" with respect to the flow of the main river. In other words, the barbed tributary joins the main river in boathook bends, which point upriver. Most barbed tributaries are the result of reversals of river flow that are caused by river piracy.

The geological story of rivers in Utah includes episodes of river capture, superposition on older structures, antecedence, drainage reversals, indifference to structure, and rivers acting to cause structure.

The Mountains
from which Rivers Flow

Rivers in Utah flow from mountains to desert basins or to the ocean. Utah's ancient rivers also flowed from mountains, but the ancient mountains, like the ancient rivers, are now gone and even the oceans have been rearranged.

Many mountain-building events of different ages have been involved in formation of the mountains of Utah. The events extend back in time to the beginning of the North American continent and beyond to the fragments of continent that were assembled to form North America. Some of the present-day mountains are youthful, and uplift continues today forming mountainous landforms that generate the source regions of rivers. Bedrock exposures record once lofty mountains that were eroded away by rivers eons ago. The oldest mountain-building events preserved in the rocks predate the assembly of the North American continent.

Mountain building is one response to the ever-changing distribution of continents and ocean basins on earth. The geological theory that applies to this process is known as *plate tectonics*, or in earlier times as continental drift. Laurentia, the Precambrian core of North America, has been a part of two supercontinents and was involved in a number of mountain-building events.

Precambrian crystalline basement in Utah represents the roots of ancient mountains, but the mountainous topography, river deposits, and landscapes have disappeared. Mountains were being made long before the North American continent took its present shape. The oldest episode of mountain building is recorded in the roots of mountains that are preserved in the Precambrian metamorphic rocks exposed in the Wasatch and eastern Uinta mountains. Additional mountains must have been produced when the crystalline basement of central Utah collided with and became a part of the older continental fragment of Wyoming and northern Utah, but those mountains have long since been eroded away. Rivers flowed southward from these mountains and their deposits are preserved in Precambrian sedimentary deposits in the Uinta and Wasatch mountains.

The western margin of the North American continent developed in latest Precambrian (Proterozoic era 900 to 543 million years ago) following separation from the supercontinent of Rodinia, and persisted to latest Devonian period (370 to 354 million years ago). A nearly continuous succession of Proterozoic to Upper Devonian sedimentary rocks accumulated in the western United States along this continental margin. The sedimentary rocks include shoreline sandstones, and carbonates that accumulated on the continental shelf and deeper water shale. The distribution of these sedimentary sequences has been severely altered by later mountain building that shortened and thickened the crust, and thrust the western sedimentary rocks over eastern continental shelf sedimentary rocks.

For 200 million years from geological periods Cambrian (543 million years ago) to Ordovician through Silurian and most of Devonian time (354 million years ago), the western continental margin was a passive continental margin. No mountains were formed on the developing western continental margin throughout the early part of Paleozoic time, and Utah was generally covered by seawater and accumulated marine sediments. Thus, no rivers or river deposits are represented in Utah during this time. But then the western continental margin became a convergent lithospheric plate boundary in Mid-Devonian (about 390 million years ago) to Mississippian time (354 to 323 million years a go). The first mountain-building event in the Paleozoic resulted in the addition of an island arc of volcanoes known as the Antler island arc. This is the first in a sequence of eight mountain-building events. This event took place far to the west in Nevada, and the faulting and other mountain-building effects are not exposed in Utah. Rivers that flowed eastward from the crest of the Antler Mountains deposited their sediment in the ocean that intervened between the Antler Mountains and Utah (Map 2). Much of Utah was part of the continental shelf where limestone accumulated.

In Pennsylvanian time (323 to 290 million years ago), major mountain building resulted from events far away from Utah on the southeastern margin of North America. The supercontinent of Pangea was being assembled from fragments of the former Rodinia supercontinent. The continent that we now know as South America became attached to North America in a collisional event that produced a major mountain chain on the southeast coast of North America and the Ancestral Rocky Mountains in Colorado. The Uncompahgre Uplift (a Ute name that means "red water springs," named for hot mineral springs near Ridgway, Colorado [Auerbach 1941]) in southeastern Utah is part of the Ancestral Rockies first elevated in late

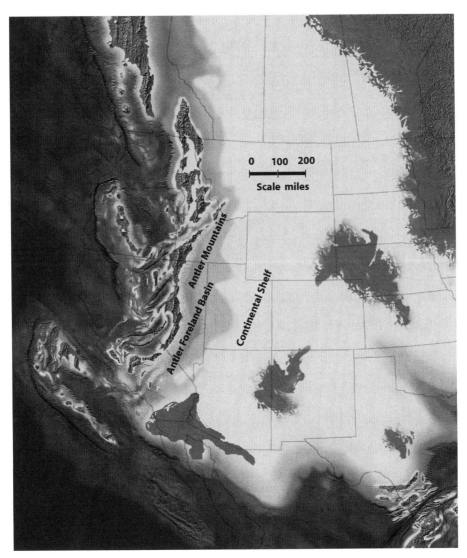

Map 2. Paleogeography of western North America in Early Mississippian time (~340 million years ago) showing the development of the Antler Mountains. Image courtesy of Ron Blakey, Department of Geology, Northern Arizona University.

Mississippian (323 to 342 million years ago) or early Pennsylvanian (323 to 303 million years ago), but uplift continued into the Triassic (248 to 206 million years ago). These mountains are shown on Maps 3 and 4.

This event also resulted in the formation of the Oquirrh Basin that covered much of northwestern Utah and the Paradox Basin that covered southeastern Utah. Both of these basins are fault block basins. As these basins subsided, they filled with sediment supplied by rivers flowing from adjacent highlands. The basin filling more or less kept pace with the subsidence.

More than 20,000 feet of sediment accumulated in the Oquirrh Basin. The Paradox Basin was less deep and accumulated less sediment. The Paradox Basin was more restricted to the circulation of the open ocean, and evaporating seawater deposited a thick layer of sodium chloride and other associated salts in the Paradox Salt Formation. The Paradox Basin was filled partially with sediment eroded from the adjacent Ancestral Rocky Mountains in Colorado. The rivers flowed from the Uncompahgre Uplift into the Paradox Basin.

The next mountain-building event began to form in Permian time (290 to 248 million years ago) and continued on into the Triassic (248 to 206 million years ago). The Oquirrh and Paradox basins had filled. Sand was blowing around on top of the Paradox Basin sediments, and the wind-blown sand formed sedimentary strata including the DeChelly Sandstone now exposed in the monuments we see in Monument Valley. Rivers that helped fill the Paradox Basin and provided some of the wind-blown sand flowed from the Uncompahgre Uplift. The western margin of North America was still a convergent lithospheric plate margin and it accreted the terrane in Nevada that we know as Sonomia. The eastern edge of accreted Sonomia is a thrust fault wedge representing the portion of Sonomia that was pushed up onto the continental shelf of North America, thus thickening the crust there. The Sonoman mountain-building event together with accreted terrane and faulting are all located in Nevada. No rivers flowing from Sonomia reached Utah, but rivers flowing westward from North America deposited sediment in a broad alluvial plain across much of Utah (Map 3).

Three Mesozoic- to Tertiary-age events are next in the sequence of mountain building. Prior to Mesozoic and Cenozoic deformation, Archean basement older than 2,500 million years was overlain by a thick sequence of continental shelf sediments of Upper Proterozoic (900 to 543 million years old) to Lower Cambrian (543 to 512 million years old) clastics and Early Paleozoic (543 to 443 million years old) carbonates. The mountain building affected great thicknesses of Early Paleozoic limestone, shale, and sandstone of western Utah. Mountain building thrust the sediments eastward shortening and thickening the crust and elevating a great mountain range. Rivers flowed eastward across Utah from this mountain belt and left behind their deposits of sandstone and conglomerate derived from the bedrock of the mountains to record the presence of both mountains and rivers.

The series of three episodes of mountain building began in the Early Jurassic (206 to 180 million years ago). These episodes of mountain building are called the Cordilleran series. The effects moved eastward through Nevada in Jurassic time (206 to 144 million years ago) and into Utah in Cretaceous time (144 to 65 million years ago). Utah was most affected in the

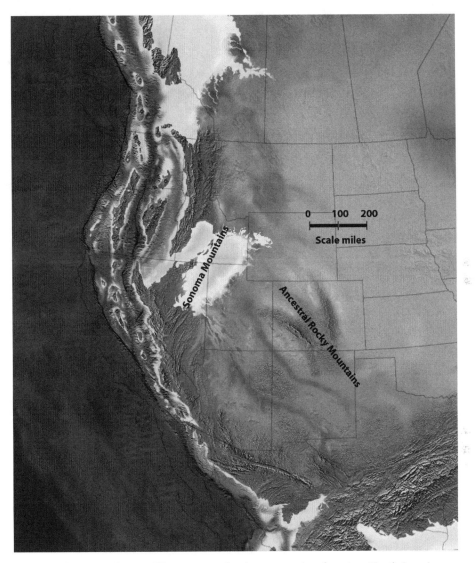

Map 3. Early Triassic (~240 million years ago) paleogeography of western North America showing the Sonoma Mountains and the Ancestral Rocky Mountains. Image courtesy of Ron Blakey, Department of Geology, Northern Arizona University.

Cretaceous and was bypassed for Colorado in Late Cretaceous (99 to 65 million years ago) to Tertiary time (65 to 34 million years ago). These three episodes produced mountains that were the sources of eastward-flowing rivers that left deposits across Utah (Map 5).

The three Cordilleran events that took place from Jurassic to Tertiary time produced a series of thrust-faulted mountain belts in Nevada and Utah. A typical mountain belt consists of, from west to east, a thrust belt, a foredeep basin, a forebulge high, and a back-bulge basin. The thrust belt

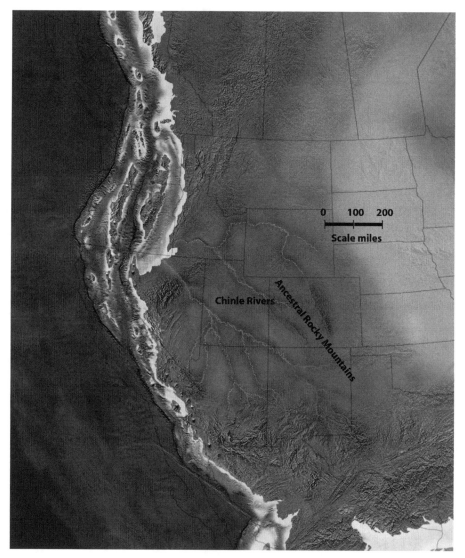

Map 4. Paleogeography of western North America in the Late Triassic (~220 million years ago) showing the course of the Chinle Rivers. Image courtesy of Ron Blakey, Department of Geology, Northern Arizona University.

consists of a thick wedge of stacked thrust plates. A single thrust plate may be up to 50,000 feet thick, and thick stacks would have formed mountains similar to the Andes. The enormous weight of the stacked thrust sheets caused the crust to sag and form the foredeep basin that could accumulate more than 10,000 feet of sediment. Rivers flowing from the mountainous stack of thrust sheets deposited sediment at the foot of the mountains as alluvial fans and braided river deposits, and carried sediment into the foredeep basin.

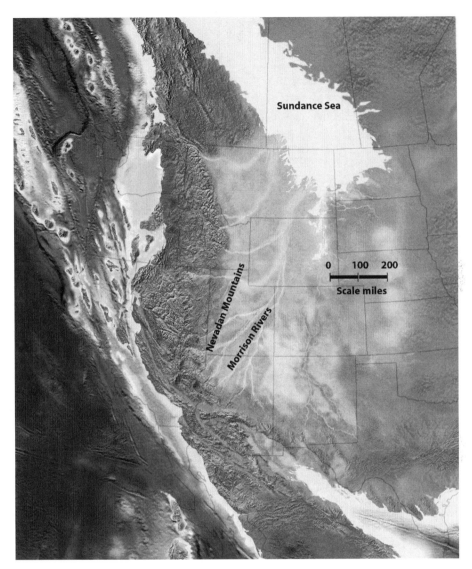

Map 5. Paleogeography of western North America in the Late Jurassic (~150 million years ago) showing the Nevadan Mountains and the course of the Morrison rivers. Image courtesy of Ron Blakey, Department of Geology, Northern Arizona University.

As the mountain-building events migrated eastward, the thrust belt, foredeep basin, forebulge high, and back-bulge basin also migrated eastward along with the rivers and their respective deposits. By Late Jurassic (159 to 144 million years ago), the back-bulge basin had migrated east of Utah into Colorado, and Utah was mostly a forebulge high. Erosion of the forebulge high left a gap in the sedimentary record of Late Jurassic river deposits between the Morrison and the overlying Cedar Mountain and Kelvin formations.

The first mountain building of the Cordilleran is a Jurassic (208 to 144 million years ago) event that produced many of the features of the Sierra Nevada Mountains, so it is called the Nevadan mountain-building event. The present topography of the Sierra Nevada is the result of later uplift and sculpting by running water and glaciers. The Nevadan mountain building formed the structure of the Sierra Nevada, but the uplift apparent today occurred much later in Pliocene-Pleistocene time (5.1 million to 10,000 years ago). During the Jurassic, rivers flowed eastward from these mountains across Utah, depositing river deposits of the Morrison Formation that is widely distributed across the western United States and is well exposed along the rivers on the Colorado Plateau (Map 5).

The Western Interior Seaway that ultimately connected the Gulf of Mexico with the Gulf of Alaska during the Cretaceous (144 to 65 million years ago) first formed in Jurassic time (206 to 144 million years ago). The compression and weight of the thickened crust of the Jurassic mountains west of Utah formed a foreland marine basin. Deposits in the Middle Jurassic (180 to 159 million years ago) back-bulge basin define a broad, shallow basin in Utah. The sequence of sediments in the basin includes deposits accumulated in a shallow ocean, on tidal flats, in flat evaporating pans, or in coastal sand dunes represented by the Twin Creek Limestone and Preuss Formation in northern Utah, the Arapien and Twist Gulch formations in central Utah, and the Entrada, Curtis, and Summerville formations in southeastern Utah. The Jurassic back-bulge basin was later covered by flood plains near sea level of the Stump and Morrison formations.

Rivers were flowing eastward across Utah into the Sundance Sea from the Nevadan Mountains out to the west (Map 5). At times, Utah was a part of the shoreline of the Jurassic Sundance Sea, and at times it was under water. Some of the sediments in Utah in Jurassic time are not marine sediments; instead, they are sediments that accumulated on land. For example, the extensive Jurassic age Nugget Sandstone in northern Utah, equivalent Navajo Sandstone in southern Utah, and the Aztec Sandstone in southern Nevada are windblown sands that accumulated next to the Sundance Sea. The Morrison Formation, exposed in Dinosaur National Monument near Vernal, consists of coarse gravel and sand that came from the Nevadan Mountains in rivers that flowed eastward into the Sundance Sea. The Morrison Formation also contains volcanic ash, evidence of volcanic activity.

Mountain building progressed across Nevada to the Eureka–Central Nevada mountain belt, and formed the thrust faults and folds of the Late Jurassic (159 to 144 million years ago); it progressed eastward and reached Utah with major effects in Cretaceous time (Map 6). Thrust faulting in Utah began in earnest in latest Jurassic to earliest Cretaceous and reached its

Map 6. Paleogeography of western North America in the Late Cretaceous (~75 million years ago) showing the Sevier Mountains and the course of the Cretaceous rivers. Image courtesy of Ron Blakey, Department of Geology, Northern Arizona University.

zenith during Late Cretaceous (99 to 65 million years ago). The thrust fault sheets are composed mostly of the sedimentary rocks that had previously accumulated on the Precambrian basement rocks. Many plates of sedimentary layered rocks in the sequence of up to five stacked thrust sheets were pushed eastward 25 to 30 miles and sometimes up to 50 miles. The thrust plates were extensively folded and faulted as can be seen in the Wasatch Mountains. Rivers that flowed from these mountains left large deposits of conglomerate at the foot of the mountains.

The thrust faulting continued into middle to late Eocene (49 to 34 million years ago), declined gradually, and ended by 40 million years ago (Lawton et al. 1997). The thrusting overlapped the Laramide mountain-building event, the seventh in the series.

The main distinction between the Sevier and Laramide mountain-building events is the thickness of crust affected. The Sevier event affected mostly the thinner sedimentary cover while the Laramide event penetrated deeper into the Precambrian crystalline basement rocks.

The Sevier event is associated with a series of thrust faults in Utah shown on Figure 4. They began with the Willard and the Charleston-Nebo thrust faults. The associated faults and folding are prominently displayed in the Wasatch Mountains, Pahvant Mountains, and other mountain ranges to the south and west.

The Cretaceous mountain-building event did not generally involve deep levels of the Precambrian crystalline basement. Paleozoic and Mesozoic sedimentary rocks were folded and faulted, but Precambrian rocks were not affected very much. The Willard thrust sheet exposed near Willard Peak is in the Paleozoic sediments. Precambrian basement rocks are prominently exposed in the Wasatch Mountains there because of the displacement on the Wasatch fault, but the Willard thrust is higher up in Paleozoic rocks. The Charleston-Nebo thrust exposed between Deer Creek Reservoir and Mount Nebo also affects Paleozoic and Mesozoic rocks. Therefore the Sevier is referred to as a thin-skinned event. The mountain range produced after the Jurassic mountain range in Nevada preserved the western interior seaway. Deposits of volcanic ash accumulated in the seaway from the volcanic activity that accompanied this mountain-building event.

The Cretaceous Sevier event was closely followed by the Laramide mountain-building event, named for the Laramie Mountains in Wyoming (Map 7). The Laramide Cretaceous- to Tertiary-age event built mountains that were the source areas for broad aprons of conglomerate many hundreds of feet thick. The rivers that deposited these conglomerates were vigorous, eastward-flowing mountain rivers that sculpted the ancient landscape and eroded to deeper and deeper levels into the stacked thrust sheets. The Laramide event was involved in formation of the Laramie Mountains, the Wind River Mountains, and the Big Horn Mountains in Wyoming, and the Uinta Mountains in Utah. The core of the Wind River Mountains consists of Archean (older than 2,500 million years) rocks that are part of the core of the ancient Wyoming continent. Since the Laramide event affected Precambrian rocks beneath the younger sedimentary cover in each of these mountain ranges, it is called a thick-skinned event. In Utah, additional Laramide structures include the Kaibab Uplift, the Monument Uplift, the San

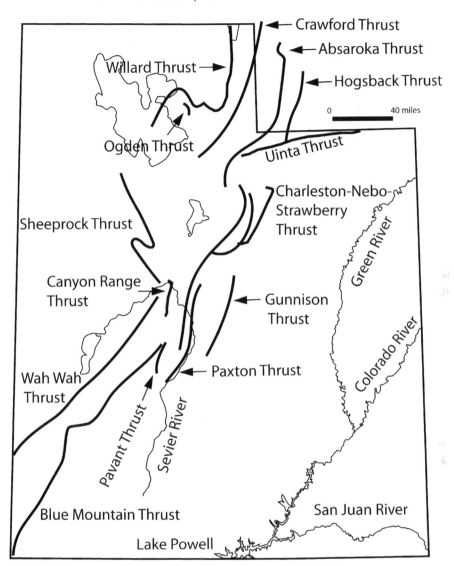

Figure 4. Late Mesozoic to Early Tertiary thrust faults in Utah. The upper plates of these thrust faults were the source of gravels deposited by ancient rivers. Locations of the thrust faults were taken from Willis (2000).

Rafael Swell, and the Waterpocket Fold (Circle Cliffs Uplift) at Capitol Reef (Map 8), all of which are cored by faults in the Precambrian basement. Vigorous mountain rivers flowed from each of these Laramide mountains and deposited thick sheets of coarse conglomerate to record their passage. Present-day rivers flow across these Laramide structures with indifference.

The thick-skinned deformation that carried eastward into the Rocky Mountains of Colorado was produced by a low-inclination or flat-dipping subducting slab of Pacific Ocean lithosphere. The flat dip was produced by

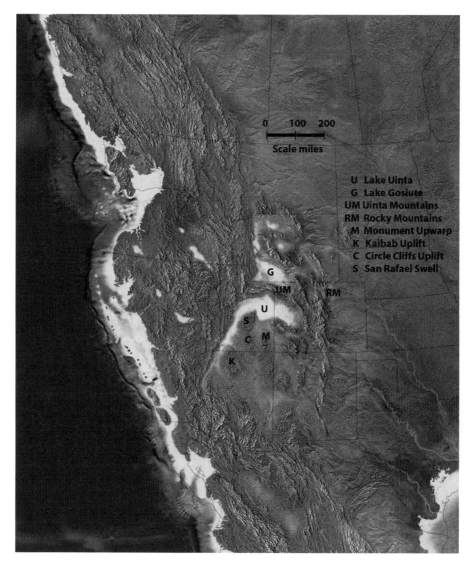

Map 7. Middle Eocene (45 million years ago) paleogeography of western North America showing Laramide-age Rocky Mountains, the Uinta Mountains, Lake Gosiute, and Lake Uinta. Image courtesy of Ron Blakey, Department of Geology, Northern Arizona University.

an increase in the rate of convergence and by subduction of younger, hotter, more buoyant oceanic lithosphere. It was not until the subducting slab reached Colorado that it was deep enough to generate igneous rocks. The flat-dipping slab also uplifted the base of the interior seaway causing the sea to recede. All the rivers continued to flow eastward from the Cretaceous Mountains located west of Utah. The rivers then flowed into large lakes that replaced the interior seaway (Map 7). Remnants of the lake deposits are preserved at Bryce Canyon and Cedar Breaks.

The final episode of mountain building is Basin and Range extension. The thickened crust produced by earlier Jurassic, Cretaceous, and Tertiary mountain-building events collapsed and began to extend westward. Constenius (1996) described a rapid drop in North America–Pacific Plate convergence rate and steepening of the oceanic slab. The result was a large reduction in east-west compression that initiated extension and collapse of the Sevier mountain wedge. Total shortening that produced the thickened crust from 150 to 50 million years ago was 65 to 85 miles. Extension due to collapse of the mountain wedge was 155 miles. The Great Basin, a region of north-south trending mountain ranges and intervening valleys with internal drainage, was formed at this time (Map 8).

Basin and Range mountain building differs remarkably from the earlier mountain-building events. The mountain building is the result of extension and normal faulting. The rocks were not extensively folded and no thrust faults were formed during this event, but some very low angle extensional faults were formed. Normal faults are responsible for the elevation of most of the present-day mountains above the valley floors.

The reasons for the collapse and extension include a thickened crust with a hot, weakened base and release of the compressive forces in the west as the San Andreas Fault formed. The western margin of North America was converted from an oblique convergent lithospheric plate margin to a transcurrent or strike-slip fault margin, the San Andreas Fault. The forces on the San Andreas Fault were insufficient to maintain the mountain elevation. The earlier formed mountains began collapsing because the force that kept the mountains elevated had been relieved. The collapse took place on north-south trending normal faults including the Wasatch fault that is still active.

Accompanying the Basin and Range mountain building was a profound change in the Colorado Plateau. The plateau began to rise and tilt to the northeast. The westernmost edge of the plateau in Utah rose to an exceptional elevation producing the High Plateaus section. As the plateau rose, rivers that flowed across the Colorado Plateau entrenched their channels. A unique pattern of deeply entrenched meanders formed on the Green, Colorado, and San Juan rivers.

Physiography of Utah

Utah is a region of mountains and valleys, plateaus and plains, canyons and rivers. All of these present-day features are the continuing result of geological processes that had their origins in the mists of the ancient past. These processes produced a basement structure and overlying geological structures that have helped determine the appearance of today's landscape. Mountain building, formation of igneous and sedimentary rocks, and landscape formation from river erosion have produced a characteristic and spectacularly scenic physiography. Three major physiographic provinces of Utah are the Great Basin, the Colorado Plateaus, and the Middle Rocky Mountains. Rivers in each of these provinces flow across characteristic and distinctive rocks and structures, and form landscapes typical of the province. Physiographic regions of Utah are shown on Map 8.

Wyoming is one big plateau broken by towering peaks. Nine peaks in the Wind River Range are over 13,000 feet in elevation, the highest being Gannett Peak at 13,804 feet. In comparison, the highest peak in Utah, Kings Peak in the Uinta Mountains, is 13,528 feet. The Green River Basin in Wyoming north of the Uinta Mountains generally slopes south with major portions 6,000 to 7,000 feet in elevation. In contrast, the central Uinta Basin south of the Uinta Mountains is 5,000 to 5,500 feet in elevation.

The Middle Rocky Mountains: The Uinta and Wasatch Mountains

The Middle Rocky Mountains in Utah consist of the Wasatch Mountains and the Uinta Mountains both composed of Precambrian, Paleozoic, Mesozoic, and Cenozoic rocks (Map 8). Most of the Rocky Mountains trend generally north and south, but the Uinta Mountains trend east and west. The Uinta Mountains consist of a large east-west-trending anticlinal uplift bounded by east-west-trending thrust faults. However, nearly north-south-trending folds and thrust faults are present in the eastern Uinta Mountains. The Uinta Mountains owe their existence to the mountain-building

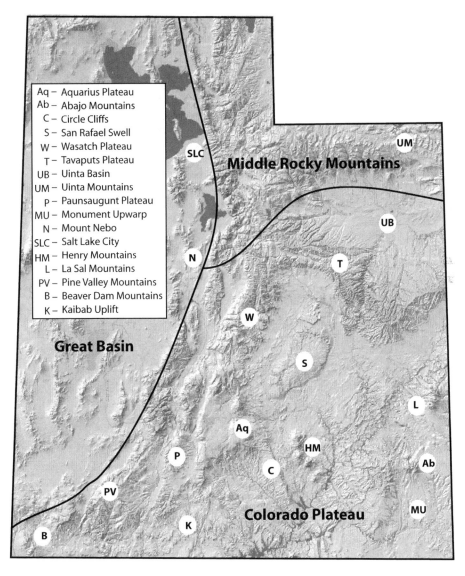

Aq – Aquarius Plateau
Ab – Abajo Mountains
C – Circle Cliffs
S – San Rafael Swell
W – Wasatch Plateau
T – Tavaputs Plateau
UB – Uinta Basin
UM – Uinta Mountains
p – Paunsaugunt Plateau
MU – Monument Upwarp
N – Mount Nebo
SLC – Salt Lake City
HM – Henry Mountains
L – La Sal Mountains
PV – Pine Valley Mountains
B – Beaver Dam Mountains
K – Kaibab Uplift

Map 8. Shaded relief map of Utah showing the location of physiographic and geographic features discussed in the text.

event known as the Laramide Orogeny that took place in late Cretaceous to Tertiary time (Map 7). The discordant east-west orientation of the Uinta Mountains results from several geological factors. The rift in which the thick, Precambrian Uinta Mountain group sediments accumulated was oriented east-west. The rigid and relatively undeformed Colorado Plateau is located to the south, and the extensive Laramide mountains of Wyoming are to the north. The Laramide mountains trend north-south in north-central

New Mexico, central and northern Colorado, and southeastern Wyoming; northwest-southeast in northwestern Colorado, central and northern Wyoming, and southernmost Montana; and east-west throughout northeastern Utah, northwestern Colorado, central Wyoming, and southwestern and central Montana. Laramide monoclinal folds in the Colorado Plateau portion of Utah trend nearly north-south.

The simplistic view of a large east-west-trending anticlinal fold such as the Uinta Mountains is that the fold is formed by north-south directed compressional forces. The Laramide mountains formed as a result of convergence between the Farallon plate in the Pacific Ocean and the North American plate. The convergence was oriented east-west producing east-west directed compressive forces that should have resulted in north-south-trending mountain structures. Work by Bruhn, Picard, and Isby (1986) and by Johnston and Yin (2001) shows that the direction of forces that formed the Uinta Mountains was directed east-northeast to west-southwest, and that transcurrent motion on the thrust faults that bound the Uinta Mountains was an important aspect of their formation. So the basic question is whether the Uinta Mountains formed from north-south compression or from east-west strike-slip faulting. New work by Johnston and Yin (2001) has resolved this issue in favor of strike-slip faulting with compressive stress consistent with adjacent Laramide structures in New Mexico, Colorado, and Wyoming.

The Uinta Mountains trend east from the junction with the Wasatch Mountains. The crest of the Uinta Mountains is broadly curved into an arc concave south toward the Uinta Basin. The range is a large compound anticline with an axis that extends west, disappears for a short distance in Kamas and Heber Valleys (Map 9), and emerges at the Wasatch Mountains as the Cottonwood Uplift where the mountain rivers of Big and Little Cottonwood Canyons flow into Salt Lake Valley. On the eastern end, the Uinta axis projects southeast to the White River Uplift in Colorado.

The Cottonwood Uplift in the Wasatch Mountains is separated from the Uinta Mountains by a deep, north-trending synclinal valley that is partly filled with Tertiary volcanic rocks where the community of Kamas is located. The uplift is a semicircular, east-plunging half-dome cored by Precambrian sedimentary rocks and middle Tertiary igneous rocks truncated to the west by the Wasatch fault.

The Uinta anticline proper is 160 miles long ending near Kamas where the western plunge of the anticline is well defined. The eastern plunge is also well defined a few miles west of Maybell, Colorado. The main Uinta anticline consists of two elongated domes separated by a structural swale in line with Manila-Vernal. Summits of the eastern Uinta Mountains are 4,000 to

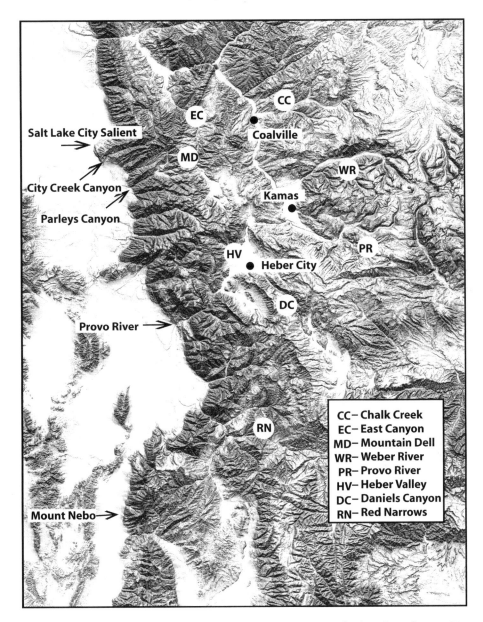

Map 9. Shaded relief map of north-central Utah showing geographic locations discussed in the text.

4,600 feet lower than the western summits. Near the east end of the anticline, the anticlinal axis has subsided and is covered by the thick Browns Park Formation.

Several smaller folds are associated with the Uinta anticline. For example, the Cross Mountain anticline straddles the Uinta axis crest and trends nearly north, and the eastern part of the range is modified on the south

33

flank by several large subsidiary anticlines centering around Dinosaur National Monument. The Uinta Range is flanked in many places by high plateaus or mesas, many capped by conglomerate. In the western part, plateaus exceed 9,800 feet, but elevation decreases to the east. Basin levels of north and south are different, but average gradients are similar from crest to base. Slopes on the south flank of the Uinta Mountains are steeper than on the north flank. The Green River Basin in Wyoming is at an elevation of 6,600 feet, while the Uinta Basin to the south is at 5,500 feet. There are ten peaks over 13,000 feet in elevation and many more over 12,000 feet.

The Uinta Mountains hosted many glaciers during the last ice age. The average glacier on the north flank was 8 to 10 miles long and two glaciers reached 20 miles in length. The average glacier on the south was 10 to 16 miles long, and six glaciers reached 20 miles in length (Stevens 1969).

The Wasatch Mountains from Nephi to the Idaho border occur at the eastern boundary of the Basin and Range where the geology changes abruptly from the folded and thrust faulted rocks of the Wasatch to the relatively undeformed rocks of the Colorado Plateau. The rugged Wasatch Mountains consist of folded and faulted Mesozoic, Paleozoic, and Precambrian rocks. The older rocks have been exhumed by erosion of the more youthful cover. Alluvial fans on the mountain flanks are faulted and merge with deltas below the level of Lake Bonneville. The northeast-trending canyons follow the sedimentary rock layers that were folded into several northeast-trending folds. The present elevation of the mountains formed from displacement on the Wasatch fault that continues today.

The Great Basin

The Great Basin is part of a larger physiographic province known as the Basin and Range province, which covers eastern Utah, Nevada, Arizona, New Mexico, Idaho, and Oregon. The Great Basin consists of many north-south-trending mountain ranges and intervening basins. In the Great Basin of Utah, there are more than sixty named mountain ranges and intervening valleys. The Great Basin appears as a region of sinuous mountain ranges generally oriented north-south with intervening valleys. The width of the valleys and mountain ranges varies. The mountains are composed of Precambrian and Paleozoic sedimentary rocks and Tertiary igneous rocks. Mesozoic sedimentary rocks are generally lacking, because the Great Basin was an area of Mesozoic mountain building and Mesozoic rocks were either not deposited there or were removed by erosion. North-south-trending normal faults form the boundaries between basins and mountain ranges. Some of the faults are active today, there are many youthful fault scarps, and the basins grow wider. Rivers leave their deposits from summer cloudburst floods as alluvial fans on the mountain flanks.

The detailed geology of each mountain range shows the presence of folds and thrust faults that were produced by compressive forces, and yet the mountain ranges are often bounded by normal faults that are the result of extension. The extended region corresponds with the region that was formerly compressed and shortened by 65 to 85 miles during earlier mountain building from 150 to 50 million years ago. The thickened crust collapsed and began to extend westward. The collapse took place on north-south-trending normal faults that are still active. Extension due to collapse of the mountain wedge is about 155 miles. The rivers that drain the mountains flow into ponds in the basins that evaporate in the summer heat, and no rivers flow out of the Great Basin into the ocean, contrary to the view of some early explorers and map makers (see Chapter 15).

The Colorado Plateau

The Colorado Plateau Province includes many mesas, plateaus, and deeply incised canyons. The general trend of elevations is dish shaped, and elevations in the center of the dish are about the same as the Great Basin to the west. Elevations in the region range from 3,000 to 14,000 feet and average 5,200 feet, about the elevation of Denver, Colorado, and 1,000 feet higher than Great Salt Lake. The Uinta Mountains form the northern boundary of the Colorado Plateau Region, and the northeastern boundary is the Rocky Mountains of Colorado. The eastern boundary is in central New Mexico, and the southern boundary is in Arizona. The western margin is a broad transition to the Basin and Range province of western Utah.

Hunt (1956) describes the Colorado Plateau as a part of the United States that is geologically better known with superb exposures of bedrock due to lack of vegetation and the deeply incised river canyons. The Colorado Plateau is about 150,000 square miles, 90 percent of which is drained by the Colorado River system, and comprises about one-half the drainage basin of the Colorado River. The Colorado Plateau may be divided into six sections: Uinta Basin, Canyon Lands, High Plateaus, Grand Canyon, Navajo (in Arizona and New Mexico), and Datil in Arizona and New Mexico. The Colorado Plateau province is roughly circular with many high plateaus and isolated mountains nearly all above 5,000 feet. Most of the sedimentary bedrock is nearly horizontal and the landscape is stepped with many cliff escarpments separated by wide gentle slopes. Geological structures present on the Colorado Plateau are broad basins such as the Uinta Basin; upwarps including the San Rafael Swell, Circle Cliffs Uplift, and Kaibab

Uplift; northwest-trending salt anticlines; northward-trending fault blocks of the High Plateaus; and intrusive domes and folds of the Henry, LaSal, and Abajo Mountains (Map 8).

The upwarps are asymmetrical anticlines with a steep eastern limb and a gently inclined western limb and have many thousands of feet of structural relief (the elevation difference in a single sedimentary bed between its position on the crest of the anticline and its position in the adjacent basin). The San Rafael Swell is a doubly plunging anticline about 90 miles long and 40 miles wide with a steep eastern limb. The Kaibab Uplift is a large, asymmetrical anticline with 5,000 feet of structural relief that extends 117 miles from northern Arizona into southern Utah. The west limb dips gently and the steeply dipping east limb, the East Kaibab Monocline, is exposed as the Cockscomb. The Circle Cliffs Uplift is 85 miles long and up to 20 miles wide. Its west limb dips gently and its east limb is the Waterpocket fold with structural relief of 5,500 feet exposed in Capitol Reef National Park. The Miners Mountain Uplift is a doubly plunging anticline 25 miles long and 15 miles wide with 5,000 feet of structural relief. The Monument Upwarp is an enormous north-south-trending anticline, 90 miles long and 35 miles wide, that extends from Kayenta, Arizona, through Monument Valley, the Goosenecks of the San Juan River, the Bears Ears, and the Needles District of Canyon Lands to the confluence of the Green and Colorado rivers. The west limb dips gently and the steeply dipping eastern limb is exposed as Comb Ridge.

This region of presumed geological simplicity is also cut by some impressive faults. Major north-south-trending normal faults on the Colorado Plateau in Utah include the Hurricane, Sevier, Paunsaugunt, and Thousand Lake faults. The Hurricane fault begins in the Grand Canyon of northern Arizona about 50 miles south of the Utah state line and extends north to disappear in volcanic rocks north of Parowan, Utah. The Hurricane fault east of Cedar City has a maximum displacement of at least 10,000 feet down to the west, and faulting puts the rim of the high Markagunt Plateau some 5,000 feet above the city. The Sevier fault begins in the Grand Canyon in Arizona and extends north 60 miles to the Utah state line and an additional 60 miles into Utah. Sevier fault displacement is at least 2,000 feet down to the west. The Paunsaugunt fault begins about 15 miles east of Monroe, Utah, and extends south to the Arizona state line. The displacement is 800 feet decreasing south to 100 feet down to the west. The Paunsaugunt fault forms the eastern boundary of the Paunsaugunt Plateau and the western edge of the Awapa Plateau. The Thousand Lake fault is east of the Paunsaugunt fault and forms the eastern boundary of the transition between Basin and Range and Colorado Plateau.

At the end of the Cretaceous, the Colorado Plateau area formed the mountain front extending east and northeast from the Sevier Mountains. Then during the Early Tertiary, it became a structural basin with interior drainage ringed by Laramide Mountains on three sides (Sevier on the west, Uinta on the north, and the Colorado Rockies on the east). No rivers flowed out of the area to the ocean, and lake deposits accumulated. Several thousand feet of river sands and gravels from the surrounding Uinta and Rocky mountains were deposited in several closed basins. The lake deposits, Flagstaff and Green River formations, record down-warping along the northwest edge of the plateau.

Then uplift began in the middle Eocene lasting until the Late Miocene, elevating the area and subjecting it to erosion, which continues today. The physiography of the Colorado Plateaus is the result of nearly continuous erosion by rivers throughout Cenozoic time that has removed thousands of feet of sedimentary cover. The western portion of the Colorado Plateaus includes a series of high plateaus. Looking westward from the highest of these plateaus is a view of the Basin and Range, the broadest zone of active continental stretching in the world. Here, the continental crust is vigorously pulled apart. Beneath this zone is a source of heat, possibly a plume of mantle material, rising from deep in the earth with a summit near the center of the Great Basin and symmetrical zones of lower elevation on either side. The heat provided by this source continues eastward beneath the high plateaus. The crust of the plateaus remains thick, as thick as it once was to the west before stretching and thinning occurred. The combination of thick crust and high heat flow has produced the elevation of the high plateaus, in excess of 10,000 feet above sea level. Yet the rocks exposed here include sedimentary rocks deposited at or near sea level.

The Gunnison and Wasatch plateaus are the northern limestone capped plateaus. The western flank of the Wasatch Plateau is a huge monoclinal fold that is nearly 50 miles long where the Tertiary sediments bend upward to the summit of the plateau. Tertiary-age volcanic rocks cap the Awapa, Sevier, Tushar, Fish Lake, Aquarius, Markagunt, and Paunsaugunt plateaus. The Table Cliffs Plateau, which could be considered the southern portion of the Aquarius Plateau, and the southern Markagunt and Paunsaugunt plateaus are topped by Tertiary-age limestone formations. Proceeding southward from the plateau tops, the sedimentary strata descend in a series of stair steps from the Tertiary pink cliffs of limestone, to the Cretaceous gray cliffs, to the Jurassic white cliffs of Navajo Sandstone, and so on. Dutton (1880) describes the view of what came to be known as the "Grand Staircase" as follows: "To the southward the profile of the country drops down by a succession of terraces formed by lower and lower formations which

come to the day light as those which overlie them are successively terminated in lines of cliffs, each formation rising gently to the southward to re-
cover a portion of the lost altitude until it is cut by its own escarpment."

Salt Anticlines

Salt deposits in the Pennsylvanian Paradox Formation and the Jurassic
Arapien Formation have produced an additional characteristic of the physiography of the Colorado Plateau and its transition with the Great Basin.
Salt beds have three characteristics that produce distinctive landforms. Salt
is less dense than surrounding and especially overlying sedimentary strata.
Salt is weak and deforms easily when force is applied. This density difference and weakness cause the salt to rise buoyantly through the overlying
strata. When the salt rises, overlying strata are folded upward into an anticlinal structure called a salt anticline. The formation of salt anticlines probably began soon after deposition of the Paradox Basin salt beds and may
continue today (Cater 1970; Doelling 1988). In addition to its buoyancy
and weakness, salt is very soluble in water. Salt in the cores of anticlines
may be over 10,000 feet thick. Dissolution of salt was the most rapid denudation process in the Paradox Basin during the late Cenozoic. Water per-
colating downward from the surface dissolves the salt core of the anticline,
causes collapse of the structure, and forms a valley. The lowest topography
is where the thickest salt has been removed and corresponds to solution collapse of salt-cored anticlines.

In Moab Valley near Moab, Utah, several thousand feet of salt dissolved
during the late Cenozoic. Overlying strata collapsed and eroded, leaving a
valley 3 miles wide and 18 miles long. Salt solution by groundwater is fa-
vored by extensional fractures along crests of anticlines and by nearby deep
canyons that produce hydraulic gradients. Removal of Triassic strata that
were barriers to groundwater flow allowed groundwater circulation into
the salt beds. Salt is still being dissolved today. Valleys formed in this way
from dissolving the Paradox Basin salt beds include Moab, Salt, Paradox,
and Fisher valleys. The axes of the anticlines, and hence the trend of the valleys, are northwesterly. The largest of the major collapsed salt anticlines are
Salt Valley, Cache Valley, and Moab-Spanish Valley anticlines. Collapse of
salt anticline crests has occurred in two stages. The first stage followed late
Cretaceous folding, and the second stage followed plateau uplift in the Tertiary. Valleys formed from collapse of Jurasssic salt-cored anticlines may
also include Sanpete and Sevier valleys, although alternative explanations
are sometimes given.

Cosmogenic Dating of Geomorphic Surfaces

The partial history of incision of modern rivers is preserved in terraces that are exposed at various locations and elevations along the course of the rivers. Several types of terraces are present. River terraces may be bedrock surfaces that have been planed off or leveled by river erosion, bedrock surfaces covered with a thin veneer of alluvium, or alluvial deposits resulting from an earlier depositional phase of river development. Terraces are exposed discontinuously along the course of most of the modern rivers described in the following chapters.

The age of various stages of river incision can now be determined using cosmogenic age determination of the surface of river terraces. The technique measures the length of time a surface has been exposed to natural cosmic rays. Primary cosmic rays are mostly protons that originate within the Milky Way galaxy. Secondary cosmic ray neutrons are generated when the primary cosmic rays strike the nuclei of atoms in the earth's atmosphere. Secondary cosmic ray neutrons continually bombard rocks on the surface of the earth. Crustal rocks are dominantly oxygen, silicon, aluminum, iron, magnesium, and calcium compounds. Cosmic ray neutrons striking these atoms dislodge nuclear particles such as neutrons, resulting in the formation of cosmogenic isotopes in the rocks. The abundance of each cosmogenic isotope is determined by the intensity of cosmic rays striking the rock and the length of time the rock has been exposed to cosmic ray bombardment. The secondary cosmic ray intensity is diminished by the atmosphere, so cosmogenic isotope production will be greater at higher altitude and higher latitude.

Helium-3 (^{3}He) is formed from fragmentation of the nuclei of oxygen, silicon, aluminum, iron, or calcium, and is usually measured in the minerals olivine or pyroxene. Helium-3 is produced in the more abundant mineral quartz but may be lost by diffusion. Beryllium-10 (^{10}Be) is produced from oxygen, silicon, magnesium, or iron and is usually measured in quartz, olivine, or magnetite. Beryllium-10 is unstable with a half-life of 1.5 million years. Aluminum-26 (^{26}Al) is produced from fragmentation of the nuclei of silicon, aluminum, or iron and is measured in quartz or olivine. Aluminum-26 is also unstable with a half-life of 700,000 years. New analytical instruments such as noble gas mass spectrometers are now capable of measuring the very low abundance of cosmogenic isotopes (Bierman 1994).

Ancient Rivers

The oldest rocks in Utah were assembled into the ancient core of the North American continent along with mountain-building events that accompanied collisions during the assembly. Rivers flowed from these mountains and deposited sediment at the base of the mountains. Just as mountains give rise to rivers today, the prehistoric mountains in Utah also gave birth to river systems. After assembly of western North America in Precambrian time, assembly of the ancient supercontinent Rodinia, and the breakup of Rodinia, mountain ranges began to form on the western margin of North America in Devonian-Mississippian time, but an ocean separated them from Utah. In Pennsylvanian time, the Ancestral Rocky Mountains formed in Colorado barely reached into eastern Utah. A large river system originated in these mountains and flowed westward through Utah, emptying into the ocean in eastern Nevada. This river system deposited sediment that is known as the Chinle Formation in southern Utah or the Ankareh Formation in northern Utah.

Later mountain-building events progressed eastward from Nevada, and the large western mountains were the source of eastward-flowing rivers that deposited sediment known as the Morrison Formation. The mountain building continued to progress eastward into Utah with still more river systems flowing eastward into the Cretaceous interior seaway. These river systems deposited sediments known as the Indianola, Frontier, Echo Canyon, and North Horn formations.

As mountain building progressed eastward into Colorado, the seaway gave way to interior lakes, with the Uinta Mountains separating a large northern lake from a large southern lake. Rivers flowing into these lakes from surrounding mountains deposited sediment known as the Wasatch Formation and the Bishop Conglomerate.

The Oldest River System

The sedimentary rocks that make up the largest part of the Uinta Mountains, the Precambrian Uinta Mountain Group, were deposited by the oldest river system known in Utah. Perhaps there are river deposits in older rocks, but metamorphism has destroyed the evidence of water flow.

The Uinta Mountain Group comprises a succession of river deposits, which have been studied by Wallace (1972) and DeGrey and Dehler (2005), among others. The oldest river system consisted of at least three separate rivers and their associated floodplains. These rivers transported sediment that consisted of quartz, feldspar, and other minerals gleaned from the exposed ancient continental nucleus in southern Wyoming, south of the Wind River Mountains and southeast of the Medicine Bow Mountains. A block-faulted mountainous complex of granite, gneiss, schist, greenstone, metavolcanics, quartzite, ironstone, sandstone, and shale supplied the rivers with a diverse mineral assemblage. The rivers, which were braided rivers merging with a delta plain, may have originated as far east as South Dakota (DeGrey and Dehler 2005). The rivers flowed south and southwest and discharged into a shallow ocean near what is now the crest of the Uinta Mountains. The sediment remaining in the river upon reaching the ocean was deposited in deltas, tidal flats, and offshore bars, and some was redistributed westward by strong longshore currents, similar to the manner in which the Mississippi River sediment is redistributed in the Gulf of Mexico today. The Red Pine Shale that covered the entire western half of the Uinta Mountains covers these deposits. This sediment was also derived from an eastern source.

No fish inhabited these rivers because fish had not yet appeared on earth. The only fossil organism that has been observed is a small, single-celled or filamentous bacterium.

Triassic River System

The Upper Triassic Chinle Formation of the Colorado Plateau and equivalent Dockum Group of eastern New Mexico and northeastern Texas consist of river sand and gravel including channels cut rather deeply into the underlying marine Moenkopi Formation. Upper layers include mudstone and siltstone members. These rocks were deposited by a river system that flowed from northern and northeastern Texas across New Mexico and southern Utah into the ocean in central Nevada where the Triassic shoreline was located (Riggs et al. 1996). The river system located near the latitude of Cuba and dating from about 225 million years ago flowed from Pennsylvanian-age highlands nearly 1,000 miles across the ancient continent of Pangea. The rivers flowed north to northwest into the ocean in what is now eastern Nevada and California, transporting distinctive minerals from the source of the river and tree trunks from the banks. Triassic rivers are shown on Map 4.

The Chinle Formation within the Circle Cliffs portion of the Grand Staircase–Escalante National Monument in Utah contains remains of one of the largest fossil forests of upper Triassic age in the world, the Wolverine Petrified Forest described by Ash (2003). Most of the petrified wood consists of wood replaced by quartz. Logs nearly 100 feet long and 1.5 to 3 feet in diameter are in the middle of the Chinle in a sandstone bed about 10 feet thick in a member of the Chinle called the Petrified Forest Member. The logs represent trees of the species *Araucarioxylon arizonicum*. A small number of *Woodworthia arizonica* are also present. Both of these species are now extinct. The trees are probably conifers of some type, but no fossil leaves, cones, or seeds have been found at Wolverine. In the Petrified Forest National Park in Arizona, *A. arizonicum* mature trees may be 10 feet in diameter 5 feet above the base and some stood 180 feet tall. *Araucarioxylon arizonicum* is the Arizona state fossil, and was given this name because the structure of the wood resembles modern araucarias (e.g., Norfolk Island pine and monkey puzzle tree) but may not be related to them. The area was a lowland, sometimes dry basin with abundant runoff from mountains to the south and west during monsoons and volcanic eruptions.

Jurassic Rivers

The landscape of Utah changed dramatically at the end of Jurassic time because of the encroaching mountain building from the west. An Andean-type mountain range along the western margin of the North American continent provided the topographic gradient and source of sediment for Morrison rivers flowing eastward across Utah and adjacent states into Colorado. River and lake deposits of the Morrison Formation covered much of Utah at the close of Jurassic time. The Morrison Formation has been studied by many hundreds of geologists because it is the host for important deposits of uranium ore; it contains abundant fossil remains of dinosaurs and other animals and plants; and it is a widespread deposit covering the eastern two-thirds of Utah, all of Wyoming and Colorado, and parts of Idaho, Montana, North Dakota, South Dakota, Nebraska, Kansas, Oklahoma, New Mexico, Arizona, and even a small corner of Texas. The late Jurassic development of the Morrison Formation ended the Middle Jurassic marine deposition in northwestern Colorado and northeastern Utah. The river deposits advanced from the southwest to the northeast as the marine shoreline regressed to the northeast in response to mountain building on the western margin of North

America. The Ancestral Rockies in Colorado and southeastern Utah that had first formed in Pennsylvanian time were no more than low hills by this time and were eventually covered by sediments of the Morrison Formation, so Morrison rocks rest directly on Precambrian rocks of the Ancestral Rockies in the Uncompahgre Uplift. In fact, reactivation of this uplift provided a source for the Recapture and Westwater Canyon members of the Morrison Formation (Dawson 1970). The Morrison Formation represents the earliest river deposits that were derived from the Mesozoic Cordilleran mountain building. The Colorado Plateau area of Morrison deposition was located at about 21° north latitude (Steiner 2003). Jurassic rivers flowing eastward are shown on Map 5.

As with many stratigraphic units, the Morrison Formation is subdivided into several members. Craig and colleagues (1955) describe these members. The lower Morrison, known as the Salt Wash Member and consisting of conglomerate, sandstone, mudstone, and claystone with a bit of limestone, is a large alluvial plain or fan formed by an aggrading system of braided rivers that diverged to the north and east from south-central Utah. The source of the rivers was southwest of south-central Utah, probably in west-central Arizona and southeastern California, because the deposits grade from coarse at the alluvial fan apex to fine at the margin. The northeast-flowing trunk rivers were mostly of low sinuosity with low braiding indices, and flow was confined between leveed banks. Tributaries to the trunk rivers were meandering rivers, and the river deposits also include crevasse splay deposits that result from flood stages breaking through the natural levees (Tyler and Ethridge 1983). The Recapture Member intertongues with the Salt Wash Member and formed as a large alluvial fan by an aggrading system of braided rivers. The Westwater Member was deposited by eastward–flowing, high-energy, intermittent rivers in a multichannel river system. The upper Morrison Westwater Member is a large alluvial plain or fan of braided rivers that flowed from a source south of Gallup, New Mexico, in west-central New Mexico. The Westwater and Recapture members are similar in distribution and both come from intrusive and extrusive igneous, metamorphic, and sedimentary sources. Finally, the upper Morrison Brushy Basin Member represents a large lake with river inputs with similar source as the Salt Wash Member. The Brushy Basin Member formed in a large, saline, alkaline lake named Lake T'oo'dichi (Navajo for Bitter Water), which encompassed both San Juan and Paradox Basins in southwestern Colorado and southeastern Utah. Lake T'oo'dichi is the largest and oldest saline lake known (Turner-Peterson 1987). The rivers that formed the Morrison beds flowed just opposite to the flow of the modern Green, San Juan, and Colorado rivers.

Rivers from Sevier Mountain Building

Later Mesozoic mountain building affected Utah in the Cretaceous. This mountain building, known as the Sevier, resulted in a mountain range that extended across Utah from the northeast to the southwest. The mountain range consisted of a series of stacked thrust sheets that formed as the mountain building progressed, which considerably shortened and thickened the crust. A foreland basin formed to the east of the mountain belt into which rivers from the mountains could flow. The thrust sheets and the resulting sedimentary deposits bear different identities and names along the thrust belt (Figure 4). In northeastern Utah, the thrust sheets are the Willard, Ogden, Crawford, Absaroka, and Hogsback. The deposits of coarse sediment that formed from these thrust sheets are the Frontier, Henefer Formation, Echo Canyon and Weber Canyon Conglomerate, Evanston Formation, and the Wasatch Formation. Paleorivers flowing eastward from the Sevier Mountains are shown on Map 6, and river deposits are summarized in Table 1.

DeCelles (1994) describes the thrust sheet deformation in northeastern Utah. The thrust sheet deformation that provided the source of rivers and the conglomerates generally progressed eastward from Willard to Ogden to Crawford to Absaroka to the Hogsback thrusts (Figure 4). The Willard is the westernmost and structurally the highest and oldest. The younger thrust faults are generally deeper and extend farther to the east. The eastern edge of each thrust sheet ramps upward to the surface, and the frontal edge is folded into ramp anticlines. The crustal shortening and resulting mountainous topography was episodic. Three main episodes of shortening produced the topographic relief from which the rivers originated (DeCelles 1994): 89–84 million years with 20 miles of shortening; 84–62 Ma (millions of years ago) with a break at 75–69 Ma that involved 18 miles of shortening; and finally 56–50 Ma on the Hogsback thrust with 13 miles of shortening. Comparison of the amount of shortening with the duration of a shortening event yields long-term rates of shortening of 0.1 inch per year to 0.25 inch per year. The three episodes of shortening resulted in a total of about 50 miles of shortening and at least 16 miles of thickening. The maximum thickness of the stacked thrust sheets is near the Wasatch Mountains. The crustal shortening and thickening produced topographic relief that is marked by three large accumulations of conglomerate totaling 10,000 feet thick.

The Henefer Formation, Echo Canyon, and Weber Canyon Conglomerate were deposited during Crawford thrusting. The Evanston Formation was deposited after the Absaroka thrust, and the lower conglomerate portion of the Wasatch Formation was deposited during and after Hogsback thrusting. Most of the sediment was derived from repeated uplift of the

Table 1. Deposits formed by ancient rivers in Utah

Formation Name	Age	Source
Morrison	Jurassic	
Brushy Basin Member	Jurassic	Lake T'oo'dichi inputs from west central Arizona and southeastern California
Westwater Canyon Member	Jurassic	west central New Mexico
Recapture Member	Jurassic	west central Arizona, southeastern California
Saltwash Member	Jurassic	west central Arizona, southeastern California
Chinle, Ankareh	Triassic	Ancestral Rocky Mountains Texas, New Mexico
Uinta Mountain Group	Proterozoic	Wyoming to South Dakota
Northern Utah		
Perrys Hollow	Miocene-Pliocene	
Evanston	Eocene (51–56 million yrs)	after Absaroka thrust
Fowkes	Eocene	
Wasatch	Eocene-Paleocene (36–38)	during and after Hogsback thrust
Frontier	Late Cretaceous Cenomanian	
Henefer	Cretaceous Sevier	Crawford thrust
Echo Canyon	Cretaceous	Crawford thrust
Weber Canyon	Cretaceous	Crawford thrust
Central Utah		
North Horn	Maastrictian-Paleocene	Charleston-Strawberry thrust
Canyon Range	Campanian (77 million yrs)	Canyon Range thrust
Price River	Campanian	
South Flat	Campanian	
Blue Castle Tongue of the Castlegate sandstone	Campanian	Last influx of thrust belt sediment
Indianola	Early Cretaceous	Charleston-Strawberry thrust
Cedar Mountain	Albian	
Southeastern Utah		
Pine Hollow	Middle Eocene	Circle Cliffs (Laramide)
Claron (basal)	Middle Eocene	Laramide uplifts
Grand Castle	Lower Paleocene	Wah Wah and Blue Mountains thrusts
Canaan Peak	Campanian to Paleocene	California, Southern Nevada
Kaiparowits	Campanian	California, Southern Nevada
Iron Springs	Santonian-Campanian	Wah Wah and Blue Mountains thrusts
Uinta Mountains, Uinta Basin		
Browns Park	Miocene	Uinta Mountains
Bishop	Oligocene (29 million yrs)	Uinta Mountains
Duchesne River	Late Eocene to Oligocene	Rising Uinta Mountains
Uinta	Late Eocene	Rising Uinta Mountains
Wasatch	Eocene	Rising Uinta Mountains, Wind River Mountains
Currant Creek Formation	Late Cretaceous	Rising western Uinta Mountains

Willard thrust sheet and eastern flank of the Wasatch Culmination. A minor source of sediment was derived from the frontal ramp anticlines. Periods of structural inactivity are marked by widespread braided river conglomerate on top of thrust belt rocks. The Henefer Formation is a fan delta front sequence of mudstone, sandstone, and conglomerate 2,600 feet thick. The Echo Canyon Conglomerate, named by Williams and Madsen (1959), contains late Cretaceous fossils and boulders up to 15 feet in diameter of Precambrian, Paleozoic, and Mesozoic rocks. Boulders decrease in size west to east as a result of the eastward direction of flow and increasing distance from the mountain front. The Echo Canyon Conglomerate, 1,700 feet thick, represents eastward-flowing rivers that deposited alluvial fans, flood and debris flows, and proximal to medial overlapping alluvial fans that prograded from the west. The Weber Canyon Conglomerate is 1,900 feet thick and was deposited by eastward-flowing rivers in East Canyon and Mountain Dell Canyon (Map 9).

The folding and faulting that produced the mountains from which the rivers flowed exposed a great variety of strata to erosion by the eastward-flowing mountain rivers. Thus the conglomerates contain fragments of all of these strata generally in inverse order as erosion proceeded to deeper and deeper levels in the thrust sheets. The Henefer contains fragments of Gartra, Ankareh, Nugget, Twin Creek, Preuss, Kelvin, Frontier, Weber, and chert from the Park City Formation. The lower Echo Conglomerate contains Weber, Park City, Nugget, Twin Creek, Preuss, Paleozoic limestones, and quartzite from Proterozoic-age rocks deeper into the Willard thrust sheet. The upper Echo Canyon Conglomerate contains Weber Formation as the vast majority of sandstone clasts, but Park City, Ankareh, Nugget, Twin Creek, and Preuss Formations are also represented. The Echo Conglomerate was derived from three separate stratigraphic sections exposed on the inactive Willard thrust sheet and eastern flank of the Wasatch Culmination (DeCelles 1988). The Wasatch Formation contains Proterozoic quartzite derived from the Willard thrust sheet. As erosion cuts deeper and deeper into the thrust sheets, older and older rocks are exposed, thus the younger conglomerates contain fragments of older rocks.

Bryant, Nichols, and Cobban (1982) have also identified three pulses of uplift in the northern portion of the overthrust belt that differ somewhat in age from DeCelles (1994). Westward thickening conglomerate tongues indicate the pulses during deposition of the Frontier Formation in its westernmost exposures in the Parleys Canyon syncline (Map 9). Uplifts are not far west or northwest of the depositional site of the Frontier. The first pulse occurred sometime between 99.6 and 89.3 million years ago and produced a tongue of conglomerate that extended as far as East Canyon Creek (Map 9).

The second uplift episode occurred between 89.3 and 85.8 million years ago and spread conglomerate nearly 20 miles farther east to Coalville (Map 9). The third pulse of uplift occurred soon after the second, and produced a conglomerate tongue intermediate in extent to the other two, extending to approximately Wanship. The conglomerates thicken westward rapidly to 6,500 to 11,000 feet at the westernmost exposures. Most of the fragments are sandstone with fewer limestone clasts derived from Mesozoic to late Paleozoic rocks. The final withdrawal of the sea that occupied the foreland basin and deposition of Henefer Formation and Echo Canyon Conglomerate occurred between 85 and 83 million years ago.

On the Salt Lake City salient (Map 9), where the Wasatch Mountains extend to the west behind the Utah state capitol building and City Creek Canyon, Mann (1974) describes the sequence of river deposits beginning with the Echo Canyon Conglomerate at the bottom, the Evanston Formation, the Wasatch Formation, the Fowkes Formation, Perrys Hollow Conglomerate, and the overlying Norwood Tuff. The Evanston Formation on the Salt Lake City Salient contains a basal white conglomerate overlain by red and tan fluvial sandstone, siltstone, conglomerate, and lacustrine limestone all beneath the City Creek Canyon volcanic rocks that provide early Eocene K-Ar ages 50.9 ± 2.0 m.y. (million years) and 55.7 ± 2.6 m.y. The age of the Wasatch Formation is constrained by the age of overlying Norwood Tuff at 36 to 38.3 million years, and the Wasatch overlies City Creek Canyon volcanic rocks providing a maximum age.

Perrys Hollow Conglomerate is an alluvial fan deposit south of City Creek Canyon, and the age is Miocene or Pliocene. Perrys Hollow Conglomerate is debris from erosion of exposures of the older Wasatch Formation. The Echo Canyon Conglomerate was deposited by rivers flowing toward the east-southeast with an eastward change from fluvial to near shore marine environment. Evanston Formation has no dominant paleocurrent direction, but the clasts are similar to Wasatch Mountain rocks, not Uinta Mountain rocks. Wasatch Conglomerate was deposited by rivers flowing from west-northwest toward east-southeast and the conglomerate thins eastward. In City Creek Canyon, the flow direction is from southeast to northwest as though the source was the Big Cottonwood Uplift.

Bryant (1990) describes the Wasatch Formation as Eocene and Paleocene sandstone, siltstone, claystone, conglomerate, and thin marine limestone. Pebbles and boulders are from diverse sources. Maximum thickness is 5,000 feet in the Mountain Dell–Porterville area (Map 9), and 4,000 feet north of the Uinta Mountains. In the Chalk Creek area (Map 9), sandstone and siltstone are overlain by interfingers of coarse conglomerate to the south derived from Paleozoic and Precambrian rocks of the Uinta Uplift.

The lower 600 to 1,000 feet is late Paleocene (61 to 55 million years old). Conglomerate in City Creek Canyon is Oligocene-Eocene (34 to 55 million years old) in age and contains boulder, cobble, pebble, and sandstone fragments. The Nugget Sandstone is the most common and conspicuous clast present, but a few volcanic lahar and tuff beds are also present. The accumulated thickness of the conglomerate in City Creek Canyon is a maximum of about 1,000 feet.

Conglomerates that have formed from other uplifts have also been called the Wasatch Formation. Conglomerate in basal and medial Wasatch Formation on the Uinta north flank was derived from the rising Uinta Mountains (Hansen 1965); clasts are Mesozoic and Paleozoic sediments exposed in the mountains and 5 to 10 percent quartzite from the Precambrian Uinta Mountain group. Wasatch Formation conglomerates in the Cumberland Gap area of Wyoming represent floodplain and delta deposits derived from both the Wind River Mountains and the Uinta Mountains (Lawrence 1963). The Wasatch Formation shows an inverse stratigraphy with younger rock fragments in the lower part and fragments from older rocks in the upper part representing the denudation of the rising Uinta Mountains.

In central Utah, stacked thrust sheets provided the topographic elevation to generate rivers and their associated deposits in the foreland basin. Tectonic loading of the thrusts produced a foreland basin and a correlative foreland bulge. Four stacked thrust plates are known as the Canyon Range, Pavant, Gunnison, and Paxton thrusts (Figure 4). Five major stratigraphic units of river deposits are the Cedar Mountain, Indianola, South Flat, Price River, and North Horn formations (Table 1) (Lawton et al. 1997). The Canyon Range Conglomerate is equivalent to all or part of these five. The Pahvant thrust fault is Late Albian (112 to 99 million years) to Early Campanian (83.5 to 71.3 million years). The Canyon Range thrust is Middle Campanian (77 million years); the Gunnison thrust is Late Maastrichtian (71 to 65 million years); and the latest (Wasatch or Paxton) is late Paleocene (61 to 55 million years) (Villien and Kligfield 1986). The Canyon Range Conglomerate was deposited by rivers flowing eastward from the upper plate of the Canyon Range thrust fault. The upper plate formed a mountain range that was uplifted by the underlying and younger Pahvant thrust fault (Lawton et al. 1997). The conglomerate is deformed, the top is thrust over the bottom part, and deposition of the conglomerate was contemporaneous with continued emplacement of the thrust sheet. The rivers were present before, and thus predated, the completion of thrust sheet emplacement. The rivers depositing the Canyon Range Conglomerate are therefore antecedent rivers.

In the Central Utah Book and Roan Cliffs Area, Francyk and colleagues (1990) describe easterly and northeasterly flowing rivers that deposited abundant quartz and rock fragments from the thrust belt to the west, producing a coastal plain and a shoreline prograding across eastern Utah. The last major influx of thrust-belt sediment is known as the Bluecastle Tongue of the Castlegate Sandstone. The sediment source then shifted to the south and southwest of the thrust belt where sedimentary, igneous and volcanic rocks contributed sediment to northeasterly flowing rivers. The rivers formed an alluvial plain transitioning to a coastal plain as the rivers became larger nearer to the seaway.

Incipient movement of the San Rafael Uplift in late Campanian may have slowed subsidence rates and caused rivers to migrate, producing localized thick, stacked channel-fill deposits of the Tuscher Formation. Deposition ended when subsidence of the foreland ceased. After a hiatus of about 10 million years, the braided rivers flowing northwest that originated in rising Laramide uplifts to the south deposited the Dark Canyon sequence of the Wasatch Formation during late Paleocene. The river flow was parallel with and along the west flank of the Uncompahgre Uplift. An extensive distal to medial braided river alluvial plain was the result. Another gap in the sedimentary record of 6 million years or more was followed by deposition of Paleocene lacustrine and alluvial strata in the Book Cliffs.

The North Horn Formation is sandstone, siltstone, mudstone, limestone, and conglomerate, Maastrictian to Paleocene in age, east of the Wasatch Plateau. The North Horn Formation was deposited in several small foreland basins associated with final movement of the Charleston-Strawberry thrust fault (Wachtell 1988). The North Horn Formation includes a basal conglomerate, middle sandstone, upper siltstone, and limestone. The basal conglomerate was deposited in gravel-dominated braided river systems flowing down individual alluvial fans at the foot of mountains. The middle gravel and sand was deposited by a braided river system flowing down alluvial fans farther from the mountain fronts. The upper portion was deposited by sandy braided rivers flowing into shallow lakes on the distal portion of the fan.

At Red Narrows in Spanish Fork Canyon (Map 9), sandstone and quartzite boulders in the North Horn were transported to the east-southeast from a source of Mesozoic sandstone and recycled Cretaceous conglomerate located immediately to the west-northwest. The probable source is the frontal fold of the Charleston-Strawberry thrust fault (Figure 4). The middle portion of the North Horn Formation is quartzite, carbonate, and sandstone cobbles transported to the east-southeast from Paleozoic sediments in the interior of the Charleston-Strawberry thrust system and from above the Sheeprock thrust that was carried piggyback on the Charleston-Strawberry thrust fault. So, the history of North Horn deposition reflects erosion to deeper levels within the stacked thrust sheets with time. The Indianola Group is older Santonian-Campanian (Cretaceous, 85.8 to 71.3 million years ago).

The Sevier-age thrust belt trends from northeastern Utah southwestward to the southwestern corner of Utah. The Wah Wah and Blue Mountain stacked thrust sheets in southwestern Utah gave rise to river systems and river sediments in the foreland basin that records Sevier- to Laramide-style mountain building described by Goldstrand (1994) as follows. The sequence of sedimentary formations is Iron Springs, Kaiparowits, Canaan Peak, Grand Castle, Pine Hollow, and the basal Claron Formation (Table 1). The oldest Iron Springs Formation (Upper Santonian 85.8–83.5 Ma to Lower Campanian 83.5–70.6 Ma) located around the flanks of the Pine Valley Mountains and the Beaver Dam Mountains (Map 8) records river deposits on a sandy, braid plain derived from the Wah Wah and Blue Mountain thrust sheets (Figure 4). The next in the sequence is the Kaiparowits Formation (Middle to Upper Campanian) and the Canaan Peak (Upper Campanian to Paleocene [65–54.8 million years]) on the Kaiparowits Plateau, which are derived from California and southern Nevada from a meandering fluvial system. The initial Laramide deposit in latest Cretaceous to early Paleocene is a coarse braided fluvial system of the Canaan Peak Formation. Then the Grand Castle Formation of lower Paleocene age is an east-to-southeast-flowing braided river system with the same source as the Iron Springs Formation. The Grand Castle Formation onlaps eastern thrusts and is folded by Laramide structures. The Laramide upwarps controlled the Grand Castle Basin. The Pine Hollow Formation (Lower Paleocene to Middle Eocene 54.8–33.7 million years) on the Kaiparowits Plateau is Lower Paleocene to Middle Eocene and is unconformable above post-Sevier conglomerate. The Pine Hollow Formation, which was derived from the west and from the northeast associated with Laramide uplifts of the Johns Valley anticline and the Circle Cliffs Uplift, consists of small alluvial fans that grade into playa mudflats. The Claron Formation consists of fluvial, deltaic, lacustrine units, and the base of the Claron is equivalent to the Pine Hollow but in a separate basin. Laramide deformation and the rivers it generated ceased in middle Eocene.

The Green River

The Green River has beckoned adventurers and explorers for more than 150 years. Flaming Gorge (now beneath a reservoir), Lodore, Split Rock, Desolation, and Labyrinth canyons, and the connection with the Colorado River and Cataract Canyon are irresistible lures for adventure. The largest rivers in Utah are the Green River and the Colorado River. The Green River is described in terms of its source, the rocks over which it flows, the geological factors that determine its course, and the basin into which it discharges. The Bishop Conglomerate and the Gilbert Peak erosion surface on which it lies form a bridge of understanding from ancient rivers that drained the Uinta Mountains to the modern rivers that flow from the crest to the flanks and the great river, the Green, that traverses the Uinta Mountains.

The Domínguez-Escalante expedition arrived at the banks of the Green River near Jensen, Utah, on September 13, 1776, and named it El Rio de San Buenaventura (St. Goodventure). The Green River was called the Seeds-ke-dee (Crow Indian for sage hen) by Indians and trappers. It was also called the Rio Verde and Spanish River, but Ashley named it the Green River because of its color (Urbanek 1988).

William Henry Ashley (1778–1838) was an American fur trader and explorer. He began fur trading in 1822, and established a system of meeting trappers at different places (rendezvous) instead of using trading posts at fixed locations. He explored the Green River south to Utah in 1825, and in 1826 went as far as Great Salt Lake. He eventually became a congressman from Missouri serving from 1831 to 1837.

The Green River begins near Green River Lakes on the west flank of the Wind River Range, Wyoming. As the river flows into the lakes, the Dinwoody glaciers cap the range above the river to the east, and Squaretop Mountain and Granite Peak overlook the river beginnings from the west. The Green River flows northwest from the Green River Lakes and makes a looping bend around Little Sheep Mountain (10,182 feet elevation) and heads south past Klondike Hill on the west. After meeting Gypsum Creek,

51

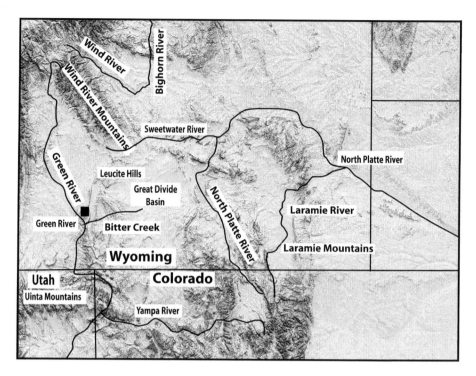

Map 10. Shaded relief map of Wyoming showing major rivers, Great Divide Basin, and the Leucite Hills. The former course of the Green River follows Bitter Creek and the North Platte River.

the Green River begins major meanders and heads south to Black Butte on the east bank. Meanders continue a bit west of south to the crossing of U.S. Highway 189/191. At Wyoming Highway 354, the river curves east in a broad valley with several channels to Daniel, Wyoming. Several miles east of Daniel, the river curves south with The Mesa on the east. Flowing south, the Green River reaches Wyoming Highway 351 and the confluence with the New Fork River, then curves west toward Big Piney and then south again paralleling U.S. 189. The Little Colorado Desert is on the east. The river flows south to LaBarge and into Fontanelle Reservoir, then drains southeast out of the reservoir to the confluence with the Big Sandy River. The river then flows southeast to Green River, Wyoming, located on Interstate 80.

At Green River, Wyoming, the Green River is joined by a long tributary in a broad valley known as Bitter Creek that extends to the east toward the continental divide. Bitter Creek may be the former course of the Green River when it flowed into the Platte River and the Mississippi River system. Geomorphic evidence, fish, and pebbles suggest that the ancestral Green River flowed east to the North Platte River (Map 10) and Mississippi River

52

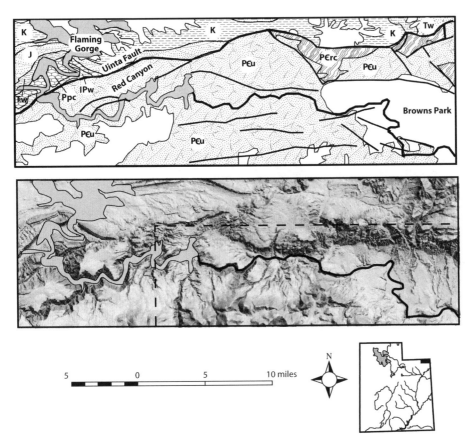

Map 11. Shaded relief and geologic maps of the Flaming Gorge–Red Canyon section of the Green River. Dashed outline shows the area of Map 12. Formation abbreviations are: Tw = Wasatch Formation, K = Cretaceous sedimentary rocks, J = Jurassic formations, Ꞁw = Woodside Shale, Ppc = Park City Formation, IPw = Weber Sandstone, Pɛu = Uinta Mountain Group, Pɛrc = Red Creek Quartzite.

drainage from early Tertiary to Quaternary (Hansen 1985). Even now, the Sweetwater River arising on the southwest side of the Wind River Range turns south then east and flows to the North Platte River.

After meandering south draining mountains and plains of southern Wyoming, the Green River cuts sharply into Cenozoic, Mesozoic, Paleozoic, and Precambrian rocks of the Uinta Mountains at the Wyoming-Utah border and flows into Flaming Gorge Reservoir (Figures 5 and 6). This section of the Green River is shown on Map 11. The river bed, now covered by reservoir waters, curves eastward because it was deflected by the Uinta Mountains through Flaming Gorge, Red Canyon, and Browns Park heading eastward into Colorado. The river flows 38 miles east following the north flank of the Uinta Mountains. In Colorado, at the Gates of Lodore, the river turns south into the steep, cliff-walled Lodore Canyon, although

Figure 5. The Flaming Gorge, a view on the Green River very near where Major Powell launched his expedition on the Colorado River. "View at the mouth of Henrys Fork (looking west) of great beauty, and which derives its principal charm from the vivid coloring. The waters of the river are of the purest emerald, with banks of sand bars of glistening white. The perpendicular bluff to the right is nearly 1,500 feet above the level of the river, and of a bright red and yellow. When entrance or gateway to the still greater wonders and grandeurs of the famous Red Canyon, that cuts its way to a depth of 3,000 feet between this point and its entrance to Browns Hole" (from the description in the U.S. Geological Survey photographic archive). U.S. Geological Survey historical photograph by W. H. Jackson.

Figure 6. Flaming Gorge looking northeast at Sheep Creek Bay and Flaming Gorge Reservoir. Author photograph.

Figure 7. Major J. W. Powell, second director of the USGS with Tau-gu, a chief of the Paiute Indians, in southern Utah ca. 1870. Published as left photo, p. 43, in USGS Earthquake Information Bulletin vol. 11, no. 2 (March–April 1979).

the broad valley of Browns Park continues eastward toward Maybell, Colorado.

John Wesley Powell (Figure 7) named Lodore Canyon on June 9, 1869, when the eighteen-year-old member of the exploration party, Andy Hall, remembered the poem "The Cataract of Lodore" by Robert Southey (Powell 1957). Soon after, one of their boats called *No Name* plunged down the rapids and encountered whirlpools and dangerous rocks. The boat struck the rocks, was broken in two, and the cargo of instruments, food, and supplies was lost. The rapids were named Disaster Falls.

The Green River Crossing of the Uinta Mountains

The Uinta Mountain landscape is a record of mountain uplift, mountain collapse, and mountain erosion. The Uinta Mountain Uplift formed in the Early Cenozoic, erosion developed a low-relief surface during the Middle

Cenozoic, and the crest of the eastern Uinta Mountains collapsed into the Browns Park graben in later Cenozoic. These geological events reversed many of the drainages and caused the Green River to traverse and incise the Uinta Uplift and join the Colorado River system (Pederson and Hadder 2005).

The Green River flows 30 miles across the Uinta Mountains in Lodore Canyon, Whirlpool Canyon, and Split Mountain Canyon before emerging from the mountains into the Uinta Basin. The Green River appears to ignore the long-held belief that rivers should flow from the crest of mountains down into adjoining basins. The Green River now flows south across the Uinta Uplift, which is opposite to the northward flow of many of its tributaries. Before reaching the Uinta Mountains, the Green River flows south across the high (6,000- to 7,000–foot) elevation Green River Basin that generally slopes south to Green River, Wyoming. The river then proceeds south and cuts deeply into the north flank of the Uinta Uplift, ignoring the Uinta Mountain structure and topography. The river then meanders in a deeply incised (1,000-foot-deep) canyon eastward to Browns Park. From Browns Park, the river enters Lodore Canyon and crosses the Uinta Uplift.

The history of ideas about how the Green River reached the Uinta Mountains and then crossed them is nearly as complex as the geological events that permitted the crossing. Pederson and Hadder (2005) have summarized the ideas on how the Green River traversed the eastern Uinta Mountains. An explanation of the incision must be consistent with the compressional and uplift history of Wyoming basins, the Uinta Mountains, the Colorado Plateau, and the Uinta Basin, and erosion of the mountains and surrounding country including the continental divide.

F. V. Hayden (1862) was perplexed by the immense canyons that he observed formed by the Madison, Jefferson, Gallatin, Yellowstone, and other rivers in the Rocky Mountains that were the sources of the Missouri River. He accounted for them by supposing that as the crests of the mountains emerged from the sea, these rivers seeking natural channels began erosive action that continued as the mountains slowly rose. The rivers were therefore older than the mountain uplifts.

John Wesley Powell offered a similar explanation for the Green River crossing the Uinta Mountains during his exploration in 1869. Powell's descent of the river in 1869 and again in 1871 produced the first accurate maps and descriptions of the rocks and gave Powell a unique perspective on the geologic setting. Powell briefly considered superposition for the origin of the valleys of the Uinta Mountains similar to the model proposed by Archibald R. Marvine in 1874. Powell supposed that Marvine's ideas worked for the main Rocky Mountains of Colorado, but on further study, Powell

Figure 8. S. F. Emmons in academic attire. Emmons was a member of the King Survey. U.S. Geological Survey photograph.

thought that the sediments exposed in the Green River Basin that he recognized as lake sediments lapped onto the Uinta Mountains and could not have completely covered them. He rejected the superposition hypothesis. Powell (1876) proposed that the course of the Green River was established before the rise of the Uinta Mountains and that the river deepened its canyon and kept pace with the uplift of the mountains. This idea is called antecedence; the river is older (antedates) the formation of the mountains. Other well-known geologists Joseph LeConte and Archibald Geike shared this idea and referred to the river as antecedent.

Powell's idea of antecedence did not stand the test of careful observation and mapping of Uinta Mountain geology. S. F. Emmons (1897) (Figure 8) of the Clarence King 40th Parallel Survey quickly realized that two large lake basins surrounded the Uinta Mountain Uplift. Rivers flowed into the lakes but did not cross them, and rivers certainly did not cross the Uinta Mountains.

Further mapping of the Uinta Mountains by Sears (1924), Bradley

(1936), and Hansen (1969) established the uplift and erosional history of the Uinta Mountains. The development of the Uinta Mountains has not been as simple as Powell envisioned. The crucial events shaping the course of the present-day Green River are related to episodes of uplift, erosion, and collapse of the mountains.

The Laramide compressional event that took place from about 60 to 30 million years ago formed the Uinta Mountain Uplift and marked the demise of the Cretaceous interior seaway. In its place, Lake Gosiute formed in Wyoming to the north of the Uinta Mountains and Lake Uinta formed in the south (Map 7). Consequent rivers drained both north and south flanks of the Uinta Mountains, depositing extensive conglomerates at the foot of the mountains and in both lakes. Lake Uinta persisted into Oligocene time as a freshwater lake that gradually filled with sediment. The last major stage of Lake Gosiute was freshwater and must have had an outlet possibly around the east end of the Uinta Mountains into Lake Uinta to the south. Both of the lakes could have drained east into the Mississippi River system. These Eocene lakes disappeared 45 to 40 million years ago. From the time Lake Gosiute north of the Uinta Mountains disappeared until the foundering of the Uinta Arch (Late Oligocene), the drainage of the southern Green River Basin in southern Wyoming was north, away from the Uinta Mountains.

The initial uplift of the Uinta Mountains along the great north and south bounding faults generated consequent streams that began the erosional history with deposition of conglomerates and sandstones. The record of uplift and erosion began with the deposition of the Currant Creek Formation, the Wasatch Formation, the Uinta Formation, and the Duchesne River Formation (Table 1). These formations contain rock fragments eroded from the rising Uinta Mountain Uplift. The lower Wasatch Formation contains fragments of Mesozoic rocks that were exposed on the flanks of the uplift, and the upper Wasatch Formation contains fragments of Paleozoic and Precambrian rocks that were gradually exposed to erosion as the uplift continued.

The early rise of the western Uinta Mountains in the late Cretaceous is indicated by the Currant Creek Formation. The Currant Creek Formation represents a braided river–alluvial fan complex that formed from southeast-flowing rivers draining the uplifted west end of the Uinta Mountains, or possibly the older Willard thrust sheet (Bruhn, Picard, and Isby 1986).

The Uinta Formation conglomerate, late Eocene in age (41.3 to 33.7 million years), is partially time equivalent to the Green River Formation that was deposited in the lakes that replaced the Cretaceous seaway. The Uinta Formation conglomerate was deposited by wide, meandering rivers that flowed west to northwest and transported debris from the Uinta Uplift (Bruhn, Picard, and Isby 1986). The Duschesne River Formation (Late

Eocene to Early Oligocene) was deposited by high-gradient braided rivers flowing south to southeast from the faulted and uplifted Uinta Mountains toward more distal lower gradient meandering rivers in the Uinta Basin. The Laramide uplift ended in the eastern Uinta Mountains after the youngest Eocene.

A long period of structural stability and erosion of the Uinta Uplift followed. Consequent rivers would have flowed north and south from the crest of this uplift into adjacent basins. There is no evidence of any steam crossing the range. The Uinta Mountains were worn down to a low-relief surface known as the Gilbert Peak erosion surface that is covered by conglomerate.

The Gilbert Peak erosion surface is a gently sloping, bedrock surface on both south and north flanks of the Uinta Mountains (Hansen 1986). When the Gilbert Peak surface formed, it was a broad, flat plain produced by river erosion (Figure 9). Toward the mountains, the erosion surface is bare rock and lacks the protective Bishop Conglomerate cover. The Gilbert surface is smoothly graded, concave upward, and slopes gently away from the mountains into the Green River Basin to the north and the Uinta Basin to the south. Rocks such as the Mississippian-age Madison Limestone rise above the Gilbert Peak surface because they are resistant to erosion. On higher flanks of the Gilbert surface on the south slope, scattered small rocky mountains that have resisted erosion merge gradually with the old, mature Gilbert Peak topography of the mountains. In such places, headward parts of the Gilbert surface penetrate deep into the range as flat-bottomed, alluviated, dentritic valleys. Many such valleys on the south slope have seen little change since the Tertiary. The Gilbert Peak surface formed in a semiarid climate shown by unweathered bedrock and lack of soil. A long period of crustal and climate stability was required to form the Gilbert Peak surface (Bradley 1936). The Gilbert surface is generally regarded as Oligocene. There are remnants of an older, less distinct erosion surface higher than the Gilbert surface called the Wild Mountain surface.

At the close of the accumulation of the Green River Formation in the lakes and the Uinta Formation, the lake basins were 500 to 1,000 feet above sea level (Hansen 1986). Formation of the Gilbert Peak erosion surface required strong uplift to produce the gradients necessary for vigorous erosion. The vigorous mountain rivers that formed the Gilbert Peak surface were likely collected into an eastward-flowing river in the Green River Basin that flowed toward the North Platte River.

Uplift of the Uinta Mountains was accompanied by regional uplift, which raised both the basins and the mountains by 3,000 feet (Hansen 1986). The center of the Green River Basin may have been at 4,000 feet

Figure 9. The Gilbert Peak erosion surface sloping northward from the south end of Little Mountain. The three photographs that make this composite were taken from Pine Mountain looking west and northwest across the valley of Red Creek. U.S. Geological Survey photograph by W. H. Bradley, from plate 36-B in Bradley (1936).

when the Gilbert Peak surface began to form and the mountains were more than 5,300 feet higher because the Gilbert Peak surface has 5,200 feet of relief and mountain peaks are 1,000 to 2,000 feet above the Gilbert Peak surface. Peaks were possibly 9,800 to 12,000 feet above sea level in Eocene time (Hansen 1986; Norris et al. 1996), although Morrill and Koch (2002) dispute the evidence for these high paleoaltitudes. The uplift rate between the end of the Eocene and Bishop Conglomerate deposition was about 3,000 feet in 9 million years (4 inches per 1,000 years to 6.5 inches per 1,000 years [Hansen 1986]). In contrast, the Laramide uplift rate that formed the Uinta Mountains was nearly 50 inches per 1,000 years.

The Gilbert Peak surface is covered by the Bishop Conglomerate deposited by consequent streams that flowed north and south from the crest of the range down the slope of the Gilbert Peak surface into the adjacent basins. The Bishop Conglomerate was named by Powell (1876:44, 169) for Bishop Mountain located just north of the Utah-Colorado line in southern Wyoming, but it has since been renamed Pine Mountain. The Bishop Con-

glomerate is widely distributed on both flanks of the Uinta Mountains, but remnants on the south flank are more extensive than those on the north. The conglomerate bed is thicker on the flanks and thins toward the Uinta Mountain crest. The Bishop Conglomerate once extended considerably farther than today, perhaps to the centers of the flanking basins, but much sediment has been removed by erosion. The source of the conglomerate and the rivers that transported it was the crest of the Uinta Mountains. As a result, the conglomerate is less coarse with distance from the Uinta crest. Sorting of the Bishop Conglomerate improves and the average size of conglomerate pebbles decreases with increasing distance from the crest of the Uinta Mountains. Gravels are also more rounded and become less coarse near the top of the conglomerate (Winkler 1970). The direction of inclination of flat pebbles indicates a river flow direction generally northward on the north flank on Diamond Mountain. Bradley (1936) also describes lenses of larger cobbles in ancient river courses that created the Gilbert Peak erosion surface.

The age of the Bishop is established by dating of volcanic beds that are incorporated in the conglomerate. Tuff in the Bishop has K-Ar ages of 29.5 ± 1.08 million years (biotite), and 28.58 ± 0.86 million years (hornblende) 295 feet above the base. The same tuff dated by Winkler (1970) had an age of 26.2 ± 0.7 million years. Two ash beds from the conglomerate on the south flank of the Uinta Mountains have been dated. The ash of Diamond Mountain Plateau has an Ar-Ar age of 30.54±0.22 million years, and the ash of Yampa Plateau has an age of 34.03±0.04 million years (Kowallis et al. 2005).

The northeast part of the Uinta fault, a great thrust fault that bounds the north flank of the eastern Uinta Mountains, was reactivated in Miocene time (23.8 to 1.8 million years ago), and the Browns Park Valley formed from collapse of the high eastern Uinta anticline along east-west-trending normal faults, causing major reorganization of the drainage (Hansen 1984, 1986).

The crest of the Uinta Mountains consisting of the oldest rocks in the core of the mountains, the Red Creek Quartzite, was lowered 5,000 feet at Browns Park as the eastern Uinta Mountains collapsed along east-west-oriented normal faults. Browns Park is a mountain valley 25 miles long and 4 to 5 miles wide along the former crest of the eastern Uinta Mountains. When Browns Park formed in the middle Tertiary, it was a canyon hundreds of feet deeper, but was later filled with sand, silt, clay, and volcanic ash of the Browns Park Formation. Rivers that had formerly flowed north and south from the crest were reversed and flowed into Browns Park Valley (Hansen 1986). The Browns Park Formation consisting of conglomerate, sandstone, and volcanic tuff accumulated in the faulted depression, and also filled valleys eroded into the Bishop Conglomerate or into the Gilbert Peak erosion surface. The source of the sediment was the Precambrian quartzite of the Uinta Mountain Group from the south and the older Red Creek Quartzite from the north.

No fossils are found in the Browns Park Formation. The oldest K-Ar date from volcanic rocks in the formation is 24.8±0.8 million years, the youngest is 11.8±.4 million years. The Browns Park Formation locally lies directly on the Bishop Conglomerate, and at other times it lies on a surface hundreds of feet below the Gilbert Peak surface (Hansen 1986).

Following Paleocene-Eocene erosion of the Uinta Mountains and deposition of conglomerate, the eastern Uinta Mountains appeared much as today, but there is a gap in the stratigraphic record. Erosion products were removed completely beyond the flanking basins. The rivers must have formed an external drainage system with rivers draining the north flank into the Platte River and the south flank into the Rio Grande. The broad valley of Bitter Creek, a tributary of the Green River that meets the river at Green River, Wyoming, may mark the course of this ancestral drainage (Map 10).

The diversion of the Green River to the south from its ancestral eastward course to the Platte River was accomplished by three events (Hansen

Figure 10. Some of Powell's party in boats in Canyon of Lodore, Green River, Dinosaur National Monument, Moffat County, Colorado, in 1871. U.S. Geological Survey photograph by J. K. Hillers.

1986). The Great Divide Basin rose and was stripped of post-Eocene rocks. The crest of the eastern Uinta Mountains subsided into Browns Park. The lavas of Leucite Hills blocked the eastern flow of the river (Map 10). Rivers flowing southward into the newly formed Browns Park vigorously lengthened their valleys northward. The result was capture of the upper Green River near Green River, Wyoming, and diversion south to the north flank of the Uinta Mountains. Following this diversion the Green River flowed eastward through Browns Park toward Maybell, Colorado, and turned southward across the Uinta Mountains to join the Colorado River system, as its course was superimposed on the Uinta Uplift at Lodore Canyon (Figure 10).

The Gilbert Peak erosion surface, the Bishop Conglomerate, and the Browns Park Formation provide a record of erosion of the Uinta Mountains by rivers that differs from the pattern of today's rivers. There is no evidence of a river that crossed the Uinta Mountains. The Green River of today must be younger than the Gilbert Peak surface, the Bishop Conglomerate, and

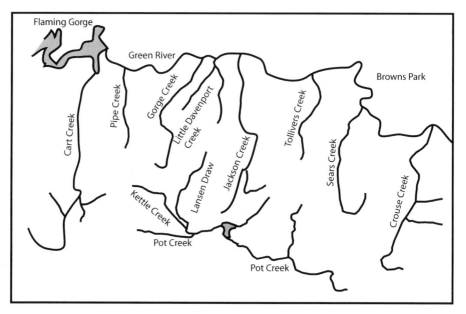

Map 12. Rivers draining the north slope of the Uinta Mountains into the Green River and Browns Park. Barbed tributaries and 180-degree bends suggest drainage reversals.

the Browns Park Formation. The Green River is therefore much younger than the Uinta Mountain Uplift.

Collapse and foundering of the Uinta crest lowered Bear Mountain by 3,200 feet or more (Hansen 1986). Subsidence of the Uinta crest into the depression of Browns Park produced a rearrangement of river drainage from Cart Creek on the west to Lodore Canyon on the east (Map 12). Consequent rivers originally flowing south from the Uinta crest over what is today Browns Park were reversed to flow north into the Green River. Evidence for this includes relict drainage divides and broad relict valleys as well as barbed-pattern headwater rivers that make 180-degree turns before joining tributaries exiting into Browns Park. Vigorous rivers that flowed into Red Canyon captured Cart, Gorge, and Jackson creeks (Hansen 1986).

Most Uinta Mountain valleys have low divides in the Bishop Conglomerate and flow north to Red Canyon or Browns Park. These rivers flowed southward from the crest of the Uintas to the Uinta Basin, but after subsidence of the crest of the Uinta arch, their courses were reversed to flow northward into Browns Park or Red Canyon. Cart Creek begins its course flowing southeast, but soon turns nearly 180 degrees to flow north, collecting tributaries that formerly joined its southward course and thus producing a typical barbed pattern. Gorge Creek, Jackson Creek, Tolivers Creek, Sears Creek, and Crouse Creek (Map 12) follow similarly reversed drainage

64

directions. The Crouse Creek drainage divide has migrated southward to nearly capture the southeastward-flowing Pot Creek. Rivers that formerly drained northward from the Uinta crest were also diverted to follow a new course southward into Browns Park. Spring, Goslin, Red, Willow, Beaver, and Vermillion creeks became the new south-flowing consequent rivers (Hansen 1986). Vermillion Creek, the largest of these with a drainage area of 1,000 square miles, formerly flowed in Irish Canyon but was captured by a tributary of the East Fork of Vermillion Creek. Pot Creek and Diamond Gulch are post-Bishop. Drainage reorganization continues today. Pot Creek is threatened with imminent capture by Crouse Creek.

Remnants of the Browns Park Formation are shown on recent geologic maps at elevations above 7,500 feet on Bear Mountain and 6,500 feet on Dutch John Bench. The Bishop Conglomerate is exposed 2,000 feet above the Green River in Lodore Canyon. Sears (1924) proposed that the Green River meanders developed on the Browns Park Formation, then were superposed onto the underlying quartzite of the Uinta Mountain Group.

These relationships suggested superposition as an alternative explanation to Powell's incorrect idea of antecedence. An ancestral Green River flowing south and east over a Browns Park surface gradually lowered itself and was superposed onto the Uinta Uplift. The Green River then crossed what had been part of the south flank of the Uinta Uplift and became part of the Colorado River system. For the Green River to be superimposed across the Uinta Mountains requires diversion of the river from its ancient course eastward to the Platte River at Green River, Wyoming, and burial of a significant portion of the eastern Uinta Mountains by Bishop Conglomerate and Browns Park Formation.

Lodore Canyon did not exist until about 5 million years ago (Hansen 1986). The Green River in Browns Park flowed southeast toward the Yampa River. The buried old Green River Valley north of the Gates of Lodore is 590 feet below the present ground surface in a canyon now filled with Browns Park Formation. Farther southeast, the canyon is buried by 1,200 feet of fill. The history of Browns Park began with the collapse of the crest of the Uinta Mountains to form the Browns Park valley. Then a valley was carved along the former crest of the Uinta Mountains. Browns Park cuts more than 3,000 feet into the quartzite core of the Uinta Mountains. The Green River was flowing sluggishly southeast in its ancestral valley in late Tertiary on the thickening fill of the Browns Park Formation. The region was gradually tilting and subsiding toward north and east, the river became overburdened with sediment, and the valley was filled with Browns Park Formation. Fill eventually overtopped the south valley rim and the river surged across the site of Lodore Canyon. The Green River was then diverted south out of Browns Park across the Uinta Mountains. The new course of the river was across the Gilbert Peak erosion surface capped by Bishop Conglomerate. The conglomerate could not have reached a level 2,000 feet above the Lodore Canyon floor if the canyon had been there first. The final phase of Browns Park history is reactivation of river erosion due to Quaternary uplift. River terraces in Browns Park mark the former stand of the river 350 feet above the present river.

Some geologists do not accept the theory of superposition alone, because it requires the Browns Park Formation to cover the eastern 50 miles of the Uinta Mountains followed by removal of the cover. On the south slope there is so much Bishop Conglomerate that there is no doubt that superposition is possible. Two additional alternative explanations have been suggested. Bradley (1936) suggested that part of canyon cutting is due to headward erosion. A river draining the south slope of the Uinta Mountains into the Uinta Basin vigorously eroded its valley headward due to the greater gradient because the Uinta Basin is 1,000 feet lower than the Green River Basin. Bradley (1936) believed that the Lodore branch of Cascade Creek (now called Pot Creek) captured the Green River. Headward erosion resulted in capture of the Green River at Lodore Canyon. Bradley agreed that the river was superposed above and below Lodore Canyon but captured at Lodore.

Hunt (1956) believes that the river is antecedent to the last stages of uplift of the Colorado Plateau. Entrenched meanders on the lower Green River clearly indicate that meanders formed on a lower gradient river floodplain, and more recent uplift of the Colorado Plateau has produced the entrenchment.

Pederson and Hadder (2005) agree that Miocene extensional faulting played a key role in diverting the Green River into Browns Park. Following the faulting, the river flowed gently through Browns Park eastward into Colorado and around the east end of the Uintas into the Yampa River. The divide of a south-flowing Yampa tributary subsided into Browns Park leaving a wind gap. Topography preserved since the Oligocene suggests the headwater drainage divide of the south-flowing paleo-Yampa also subsided into Browns Park. The Gates of Lodore follow the trend and spacing of the ancient Oligocene drainage. The pre-Browns Park drainage divide (Laramide) could have been in the heart of Browns Park near the present Gates of Lodore. Pederson and Hadder speculate that the upper Green was diverted through the fallen divide over Browns Park fill. Upper Zenobia Creek (Map 13) is a wide, flat-bottomed, upper valley that trends parallel to Lodore but perpendicular to modern creeks plunging into the canyon. Also, the head of Jack Springs Draw and the north edge of Buster Basin are wide wind-gap valleys adjacent to the Gates of Lodore. The upper cliffs of Lodore Canyon may be inherited from an old south-flowing valley.

Three mechanisms have been suggested for the Green River crossing at Lodore Canyon: headward erosion of small, south-flank rivers and capture of the Green; superposition; or part antecedent to Quaternary uplift. Superposition is a major part of the history of both the Green and Yampa rivers. Antecedence to the last uplift of the Colorado Plateau should still be part of the explanation (Hunt 1969), but Powell's idea of a river antecedent to the initial uplift of the Uinta Mountains has no support. Hansen's conclusion is that superposition was the major process when the Browns Park Formation completely filled its basin and the river spilled south at Lodore Canyon.

Browns Park to Split Mountain

Hansen (1996) described the geology along the Green River from Browns Park to Split Mountain. In this 44 mile stretch, the Green River descends 555 feet. Rates of fall vary due to rock characteristics such as hardness, fracturing habit, and dip. Rates of fall are also influenced by erosion and deposition along the riverbed. The course of the Green River across the uplift begins at Lodore Canyon (Figure 10, Map 13) in the Uinta Mountain Group sediments of Precambrian age and proceeds southward through younger rocks that include, in order, the Lodore Formation, the Madison Limestone at Echo Park where the Yampa River joins the Green in an entrenched meander bend, the Round Valley Formation as the river enters Island Park, then the Morgan Formation and the Weber Sandstone. Green River whitewater rapids in Lodore Canyon are named Disaster Falls, Hells Half Mile (greatest rate of descent 30 feet in one-half mile), and Triplet Falls.

The Yampa River joins the Green River at Echo Park. Here, the Green River makes a long, entrenched meander bend around the resistant Weber Formation that forms Steamboat Rock, crosses the Mitten Park fault, and flows into Whirlpool Canyon (a 2,500-foot-deep gorge). From Whirlpool Canyon the river meanders through Island Park and Rainbow Park and enters the gorge of the Split Mountain anticline where the river gorge is 2,800 feet deep in limestone and sandstone of the anticline core. The Weber Sandstone is exposed in high cliffs where the Green River exits Split Mountain (Figure 11). The Split Mountain anticline was formed during the Laramide uplift of the Uinta Mountains, long before the Green River reached this area. The anticline was eroded to a flat plain and covered with Bishop conglomerate and a river meandered over the plain. New uplift resulted in downcutting and the river was superimposed on the Split Mountain anticline. So, here again, Powell's idea of antecedence is incorrect.

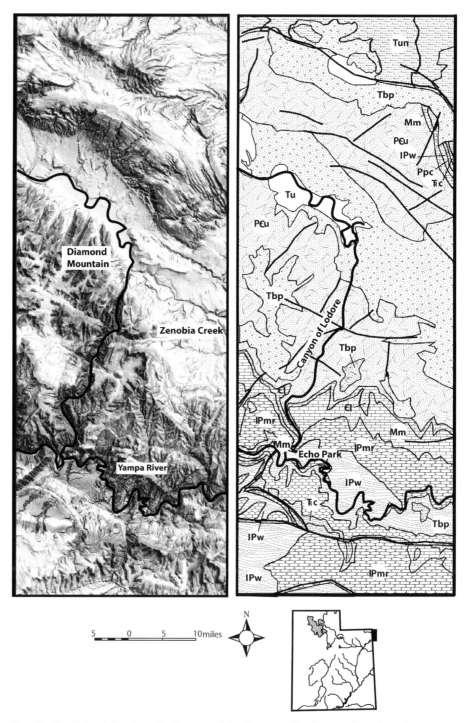

Map 13. Shaded relief and geologic maps of the Canyon of Lodore section of the Green River. Formation abbreviations are: Tbp = Browns Park Formation, Tu = Uinta Formation, Ŧc = Chinle Formation, Ppc = Park City Formation, IPw = Weber Sandstone, IPmr = Morgan and Round Valley Limestone, Mm = Madison Limestone, Єl = Lodore Formation, Pєu = Uinta Mountain Group.

Figure 11. The Split Mountain anticline near the entrance to Split Mountain Canyon on the Green River. The steeply inclined sedimentary beds are part of the Weber Formation. The river carves a canyon through the anticline. Photograph courtesy of Gary Colgan.

Split Mountain to Green River, Utah

At Split Mountain, the Green River, heading south, cuts through Park City, Weber, and Morgan Formations, then heads west into Mississippian-age rocks in the core of the anticline (Map 14). The river then heads south back through Morgan, Weber, Park City and the Mesozoic Moenkopi, Chinle, Nugget, Carmel, Entrada, Curtis, Morrison, Dakota, and Cedar Mountain, all steeply dipping on the south flank of the Split Mountain anticline.

After exiting the gorge of Split Mountain, the river cuts Mowry Shale and Frontier Sandstone in a big meander bend named Horseshoe Bend (Figure 12) repeating the Morrison, Curtis, and Cedar Mountain formations. The river makes a big meander bend to the north and then back around south to Jensen, Utah, near the Escalante Crossing where Brush Creek enters the river. The river has changed from a bedrock river in Split Mountain to an alluvial river on the Mancos Shale heading southwest into Mesa Verde Group rocks. The river is still an alluvial river meandering in Duchesne River and Uinta formations that are themselves river deposits eroded from the rising Uinta Uplift. The Duchesne River and White River enter here as the Green crosses from the river deposits of the Uinta Formation into the lake deposits of the Green River Formation (Map 15). Nine Mile Canyon comes in here as the river cuts lower into the Green River Formation and becomes a bedrock river once again.

Desolation Canyon is nearly 5,000 feet deep in the Green River Formation (Map 16). The Green River cuts through the Green River Formation into the Wasatch Formation, another of those ancient river deposits eroded from the rising uplifts. Then the Green River cuts into the North Horn Formation that formed from rivers flowing east from the Sevier thrust belt, then the river cuts into the Tuscher Formation, also conglomerate and sandstone, and then the Price River Formation (Map 17). Range Creek enters here as the river cuts through the Cretaceous Castlegate Sandstone, and then through the Blackhawk Group of the Book Cliffs, where the Price River enters. The Green River then cuts into the Mancos Shale and reaches Green River, Utah. The river has followed a fairly straight course from the Uinta Basin through the Roan Cliffs, the Tavaputs Plateau, and the Book Cliffs to Green River, Utah.

Green River, Utah, to the Colorado River Confluence

The Green River from Green River, Utah (Map 17), to the confluence with the Colorado follows a meandering course through open desert or beneath towering walls of Pennsylvanian to Late Cretaceous sedimentary strata that dip gently upriver so the course downriver is into older rocks at river level.

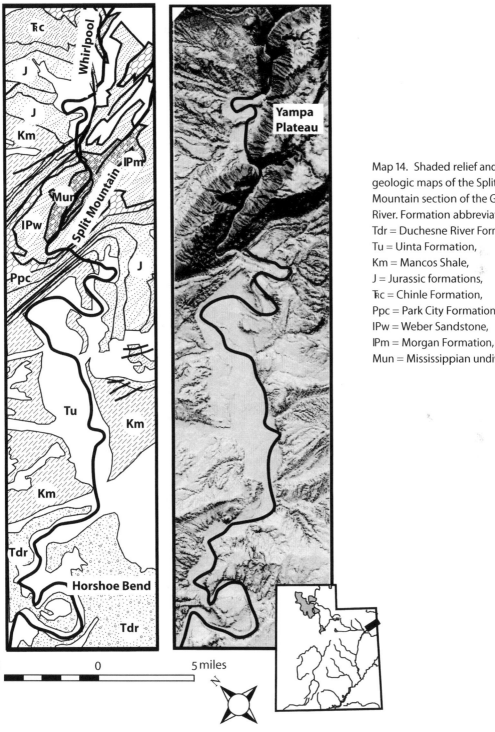

Map 14. Shaded relief and geologic maps of the Split Mountain section of the Green River. Formation abbreviations are: Tdr = Duchesne River Formation, Tu = Uinta Formation, Km = Mancos Shale, J = Jurassic formations, ℟c = Chinle Formation, Ppc = Park City Formation, IPw = Weber Sandstone, IPm = Morgan Formation, Mun = Mississippian undivided.

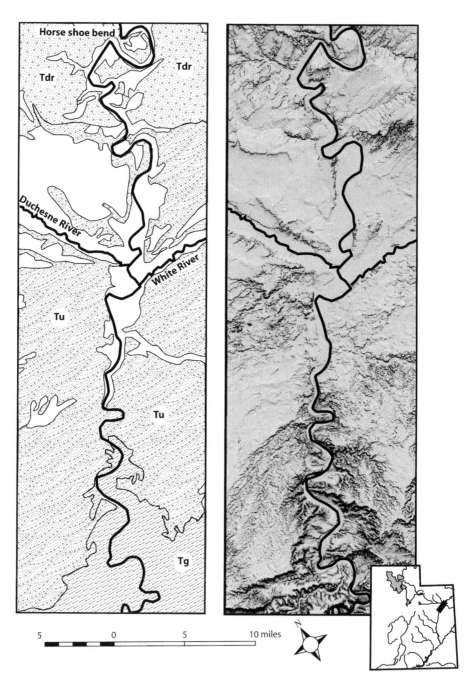

Map 15. Shaded relief and geologic maps of the Horseshoe Bend section of the Green River. Formation abbreviations are: Tdr = Duchesne River Formation, Tu = Uinta Formation, Tg = Green River Formation.

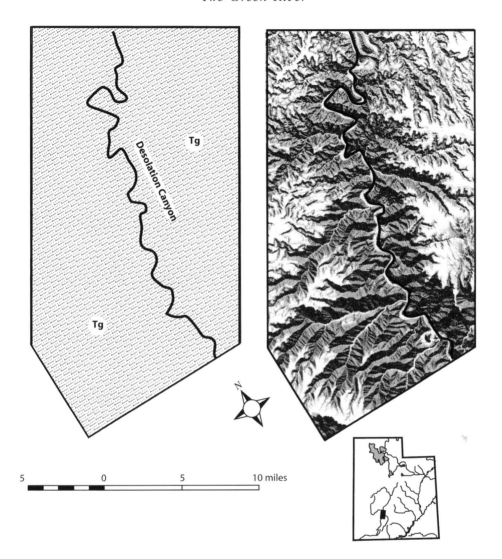

Map 16. Shaded relief and geologic maps of the Desolation Canyon section of the Green River. Formation abbreviation is: Tg = Green River Formation.

The meandering course has been maintained during the last 1,000 feet of downcutting. Green River, Utah, is a historic fording point used by Indians, Spanish, mountain men, and explorers including John W. Gunnison. From Green River, Utah, the view upriver includes the Book Cliffs and Gunnison Butte at the mouth of Gray Canyon. In Labyrinth and Still Water Canyons (Map 18) the river flows quietly, but there is a dramatic change below the junction with the Colorado River. There the walls are higher, topped by jagged spires and angular cliffs of dark Pennsylvanian-Permian rocks.

Map 17. Shaded relief and
geologic maps of the Book
Cliffs–Green River, Utah
section of the Green River.
Formation abbreviations are:
Tg = Green River Formation,
Tw = Wasatch Formation,
TKnh = North Horn Formation,
Kpr = Price River Formation,
Km = Mancos Shale.

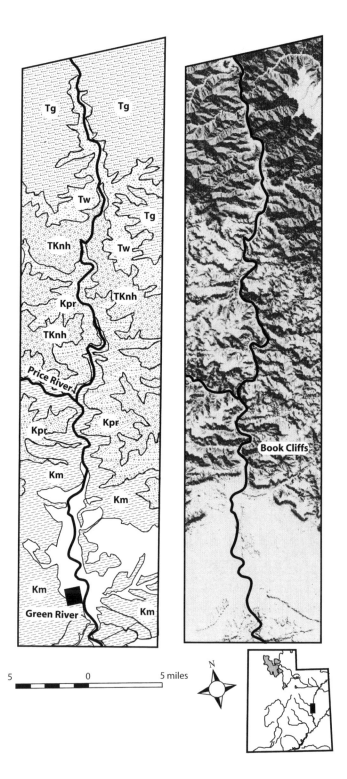

Figure 12. Horseshoe Bend and the Green River below the bend viewed from Flaming Gorge Ridge. U.S. Geological Survey photograph by T. H. O'Sullivan on the Geological Exploration of the Fortieth Parallel (King Survey). Horseshoe Bend is cut into the Tertiary-age Duchesne River Formation.

At Bowknot Bend (Map 18, Figure 13), the Green River is in Chinle Formation with a thin cliff band of Wingate and Kayenta on the surface of Spring Canyon Point and Bowknot Bend. The geology of this area is shown on the map of Huntoon, Billingsley, and Breed (1982). Navajo Sandstone is a short distance to the west. The river meanders generally southward through Cottonwood Bottom, Tidwell Bottom, and Mineral Bottom (where Mineral Canyon joins from the east), then cuts the Grays Pasture syncline in Chinle Formation with Wingate cliffs and a plateau of Kayenta above with just a few remnants of Navajo Sandstone away from the river at Bonita Bend (Figure 14). The river passes Tidwell, Horsethief, Woodruff, Point, Saddle Horse, Upheaval, Hardscrabble, Fort, and Potato Bottom, all alluvial stretches.

At Saddle Horse Bottom, the river is in Moenkopi Formation with more and more Moenkopi Formation exposures as it reaches Potato Bottom where the river is in the White Rim Sandstone Member of the Cutler Group. The river is in White Rim at Potato, Beaver, Queen Anne, and Anderson, all alluvial bottoms. At Anderson Bottom, the river cuts into the Organ Rock Shale. At Valentine Bottom, the river cuts into the Cedar Mesa Sandstone (Cutler Group), then into the Elephant Canyon Formation. The river cuts

Map 18. Shaded relief
and geologic maps of the
Labyrinth Canyon section of
the Green River. Formation
abbreviations are:
Km = Mancos Shale,
Jm = Morrison Formation,
Jscu = Summerville and
Curtis formations,
Jca = Carmel Formation,
Je = Entrada Formation,
Jna = Navajo Sandstone,
Ŧk = Kayenta Formation,
Ŧc = Chinle Formation,
Ŧm = Moenkopi Formation,
Pcu = Cutler Group.

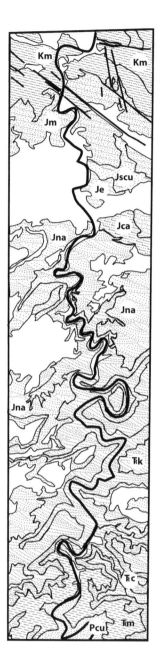

5   0   5 miles

Figure 13. Bowknot Bend on the Green River in Canyonlands National Park looking east from the west end of a narrow intervening saddle. U.S. Geological Survey photograph by E. O. Beaman, September 10, 1871, on the second expedition of the USGS (Powell Survey), Figure 62 (upper portion) in Lohman (1974). Bowknot Bend is a loop on the Green River that is 7.5 miles long. Wingate cliffs are shown above slopes of Chinle Formation. Moenkopi Formation is exposed at river level. Kayenta Formation covers the mesa. The Wingate cliffs form a band too narrow to show on the geologic map.

Figure 14. Bonita Bend on the Green River. It is here that Stillwater Canyon begins and Labyrinth Canyon ends. U.S. Geological Survey photograph by J. K. Hillers. Photograph shows Wingate Sandstone cliffs above Chinle and Moenkopi Formation slopes. Navajo Sandstone cliffs can be seen in the distance.

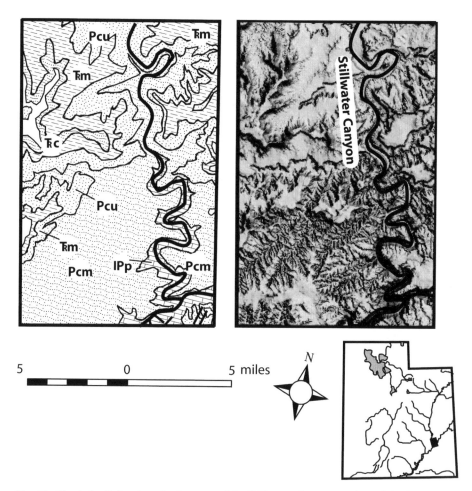

Map 19. Shaded relief and geologic maps of the Stillwater Canyon section of the Green River and the confluence with the Colorado River. Formation abbreviations are: Ŧc = Chinle Formation, Ŧm = Moenkopi Formation, Pcu = Cutler Group, Pcm = Cedar Mesa Sandstone, IP p = Paradox Formation.

into the oldest rocks since it departed Green River, Utah, the Honaker Trail Formation, part of the Pennsylvanian-age Hermosa Group that includes the Paradox Formation at its confluence with the Colorado River (Map 19). Then the Green and Colorado, both confined within deep and narrow canyon walls, flow quietly together to Cataract Canyon, where rapids dominate the character of the Colorado.

The Green River from Flaming Gorge dam to the Colorado border provides 30 miles of Utah's most famous trout fishing, and may be the most prolific trout river in the United States. The river below Flaming Gorge to Little Hole contains 20,000 rainbow, brown, cutthroat, and brook trout per mile (Demoux 2001; Prettyman 2001). Brown trout are the dominant species, rainbow trout are stocked, and cutthroat trout can be caught (Cook 2000). From the dam to Little Hole, 7 miles of crystal clear water contain many visible fish. From Little Hole to Indian Crossing in Browns Park, 9 miles of river contain 12,000 to 14,000 fish per mile. From Indian Crossing to Swallow Canyon in Browns Park, the river contains mostly brown trout with some rainbow trout. The Colorado state line marks the boundary from cold- to warm-water fish species (Demoux 2001). Below Lodore Canyon the Green River reenters Utah. From the Colorado-Utah border to the confluence with the Colorado River, the Green River can be fished for channel catfish and northern pike.

The Colorado River

The Colorado River had various names over the centuries including Rio del Tizon (Firebrand River), Buena Guia of Alarcon, Rio Grande de Buena Esperanze (River of Good Hope), River of Mystery, San Clemente (Merciful), Red River, and Pahaweep (Water Down Deep in the Earth). Colorado ('red-tinted' in Spanish) was used for the Little Colorado in 1604 by Oñate, a Spanish explorer, and was transferred to the largest river, the Grande River (Van Cott 1990).

Neither geologists nor explorers could have imagined or invented the Colorado River. The Colorado River basin is 244,000 square miles. The western Colorado River basin in south-central Utah is 15,000 square miles. The Colorado River extends into the Southern and Middle Rocky Mountains, Wyoming Basins, Colorado Plateau, and Basin and Range physiographic provinces. The river and its tributaries cross many mountains and high plateaus that are structural barriers of resistant rock. No other river in the Western Hemisphere crosses so many barriers (Hunt 1969).

The Colorado River crosses mountain barriers and the Colorado Plateau, a tremendous structural block covering more than half of the drainage basin. The Colorado Plateau has been uplifted and tilted northeast against the flow of the river. The Grand Canyon—where the Colorado River leaves the Colorado Plateau—is cut across the highest topographic and structural part of the plateau rim. Much of the Colorado River's 300-mile course across the Colorado Plateau is in steep-sided, narrow, rock-walled canyons. Broad alluvial plains are not common. The river meanders greatly, much more than most rivers that are not on alluvial floodplains, and the meanders are often prisoners of steep, narrow canyons.

The Upper Colorado above the confluence with the Green River is mostly in the Southern Rocky Mountains in Colorado and the northeast part of the Colorado Plateau in Utah. The present drainage is askew. All major tributaries enter from the south. The upper 250 miles of the Colorado River have no large tributaries from the north. South tributaries are the Blue, Eagle, Roaring Fork, Gunnison, Dolores, and San Miguel rivers. The skewness indicates this portion of the Colorado River is young, because of Quaternary (last 1.8 million years) deepening of the river canyons in the Rocky Mountains that has been estimated to be about 750 to 1,000 feet.

The Colorado River begins near the continental divide in north-central Colorado on the east side of the Never Summer Mountains. The Rabbit Ears Range lies to the west, the Mummy Range to the east. The youthful Colorado River flows south down Kawuneeche Valley parallel to U.S. Highway 34 and flows into Shadow Mountain Lake, then out of that lake into Lake Granbe, and out of Lake Granbe at Granbe Dam. The river then flows generally southwest to U.S. Highway 40 into Windy Gap Reservoir, follows U.S. 40 to Hot Sulphur Springs, then to Kremmling; west of Kremmling it flows across the Gore range at Gore Canyon. The Colorado River, heading southwest, follows the Denver and Rio Grand Railroad to Colorado Highway 131, turns northwest as it curves around Yarmony Mountain, then gradually curves southeast again still followed by the railroad. The Colorado heads southwest to Interstate 70 and there is joined by the Eagle River. The river and I-70 traverse Glenwood Canyon with the White River Plateau to the north. Immediately upriver of the Colorado Plateau in the vicinity of Glenwood Springs, Colorado, river gravels are found lying on 10-million-year-old basalt (Patton et al. 1991). These gravels predate 3,300 feet of downcutting in Glenwood Canyon.

The river cuts Grand Hogback to Rifle, Colorado, along I-70. The Colorado River, headed to Grand Junction, cuts through the Book Cliffs in De Beque Canyon, curves around Grand Junction, and heads northwest where it is joined by the Gunnison River, then cuts through Piñon Mesa at Horsethief Canyon and curves back to the southwest in Ruby Canyon and flows into Utah just south of I-70.

The Colorado River flows around the north end of the Uncompahgre Arch or Plateau where large areas exceed 9,000 feet elevation. The Uncompahgre is a northwest-plunging anticlinal uplift 25 to 30 miles wide and 100 miles long. This present-day physiographic expression of the Uncompahgre Arch has been uplifted repeatedly since its formation in Pennsylvanian time.

The Colorado River may have once crossed the Uncompahgre Uplift at Unaweep Canyon (Patton et al. 1991). Unaweep Canyon is the most spectacular diversion of the whole drainage basin. It is 2,000 feet deep, 4 miles wide at the rim, and 1 mile wide at the floor in Precambrian metamorphic rocks. More than 1,000 feet of river gravel underlie the present river divide in Unaweep Canyon. Underfit rivers too small for their valleys drain each side of the divide that heads 2,000 feet above the Colorado River. The

Southwest flank of the Uncompahgre Plateau has been uplifted 1,400 feet since Unaweep Canyon was abandoned by the river that carved it, and the Dolores River has downcut about 100 feet. In latest Miocene, the ancestral Colorado River flowed across the site of Unaweep Canyon long after it had trenched its course through Grand Mesa and volcanic fields. The rising Uncompahgre Arch forced the river to cut deeply, forming Unaweep Canyon. Meanwhile, a minor tributary north of the arch in easily eroded Mancos Shale worked headward and captured the Colorado in Pliocene time. For a time the Ancestral Gunnison River continued to flow in Unaweep Canyon, but it was unable to keep pace with the rising arch. In late Pliocene or early Pleistocene, the Gunnison River was captured by a minor river at the northeastern edge of the plateau and assumed its present course. Uplift continued into the Pleistocene, broadening the arch and extending it northwestward so the Colorado again became superimposed and carved the deep Ruby and Westwater canyons.

The Colorado River flows from Ruby Canyon in Colorado southwest into Westwater Canyon in Utah. In Ruby Canyon, high cliffs of Wingate sandstone and overlying Kayenta Formation overlook the river. In Westwater Canyon near the Colorado-Utah Border (Map 20), Triassic Chinle lies directly on the black, polished, Early Proterozoic gneisses, schists, felsic, and intermediate dikes that make up the basement of the Uncompahgre Uplift shown on the map of Gualtieri (1988). From the outcrop of the Chinle nearest the river, successive layers of Wingate, Kayenta, and Entrada lead in steps up to the Book Cliffs overlooking Grand Valley. Bitter Creek and Westwater Creek enter at Westwater where boats are launched at the head of Westwater Canyon, at an elevation of about 4,300 feet (Map 20). Westwater Canyon is a bedrock canyon with little alluvium.

Westwater Canyon to Moab, Utah

Westwater Canyon is the first whitewater stretch of the Colorado River in Utah with several challenging rapids for experienced boaters only. The first two of the eleven named rapids within Westwater Canyon are Wild Horse and Little Dolores. The most significant rapids in the 2-mile stretch of the inner canyon formed by Precambrian rock are Marble Canyon, Staircase, Big Hummer, Funnel Falls (most trouble), Surprise, Skull, Bowling Alley, Sock-It-To-Me (much trouble), and Last Chance. The river peaks at about 20,000 cubic feet per second in late May or early June (flow could be over 50,000 cubic feet per second in some years) and declines to less than 3,000 cubic feet per second. Westwater is in a broad alluvial valley 5 miles long and 1 mile wide (Map 20). Jurassic Saltwash Member of the Morrison

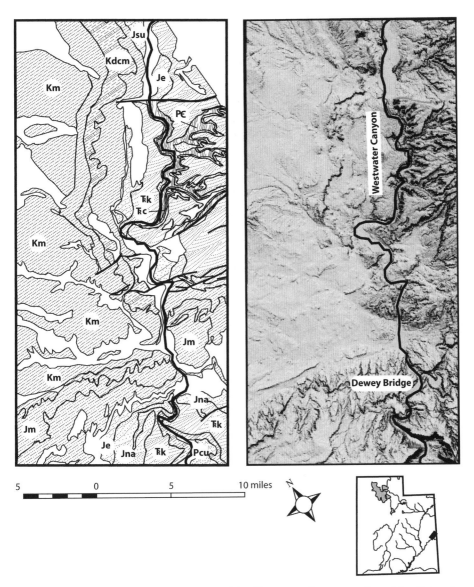

Map 20. Shaded relief and geologic maps of the Westwater Canyon section of the Colorado River. Formation abbreviations are: Km = Mancos Shale, Kdcm = Dakota Sandstone and Cedar Mountain Formation, Jm = Morrison Formation, Jsu = Summerville Formation, Je = Entrada Formation, Jna = Navajo Sandstone, Ŧk = Kayenta Formation, Ŧc = Chinle Formation, Pcu = Cutler Group.

Formation, Summerville, and Entrada appear on the northwest bank, and Kayenta, Entrada, Summerville, and Salt Wash Member of the Morrison Formation appear on the southeast bank. The river plunges into Westwater Canyon with Precambrian crystalline rocks overlain by Chinle, Wingate, and Kayenta on the southeast bank and a little less Precambrian outcrop appears on the northwest bank. Above and below Westwater Canyon, the Colorado River meanders serenely through the red rock cliffs. The river passes Little Hole, a tributary canyon in Chinle, Wingate, and Kayenta. The river emerges from Westwater Canyon in a big meander bend at Sand Flat, but just before emergence it passes Big Hole, an abandoned meander bend. The Colorado River, after the big looping meander bend at Sand Flat, crosses the west-northwest-striking Dry Gulch fault that dips south and exposes Precambrian rock on its footwall in a narrow band paralleling the fault. Then a couple of miles south at a sharper bend to the east, the river crosses the Ryan Creek fault zone. A couple of miles farther south, the river crosses the subsurface trace of the Uncompahgre fault, then is joined by the Dolores River.

The Dolores River, which begins in the San Juan Mountains of Colorado, meanders northwest from Colorado into Utah following the subsurface trace of the Uncompahgre fault and the axis of the Sages Wash syncline before joining the Colorado.

The elevation of the Colorado River as it emerges from Westwater Canyon just upriver from Hallet Ranch is 4,148 feet. The Colorado River, flowing southwestward from Precambrian basement covered with Chinle, crosses the Uncompahgre fault, covered by younger sedimentary rocks, and flows into the Paradox Basin with a thick sequence of sedimentary rocks. The sedimentary sequence includes thick salt beds that have formed salt anticlines. The Colorado River and its tributary the Dolores cut across the faulted salt anticlines. The rivers are older than, and antecedent to, the Late Tertiary to Quaternary deformation of the salt anticlines.

The Colorado then continuing southwest cuts across the structural grain of northwest-striking, salt-cored anticlines beginning first with the Fisher Valley salt-cored anticline and the Onion Creek diapir, then the northwest extension of the Cache Valley salt-cored anticline, then the Castle Valley salt-cored anticline that extends northwest from the flank of the LaSal Mountains, then the Moab salt-cored anticline (Map 21).

The history of uplift and collapse of the salt-cored anticlines is complex and paradoxical. These salt-cored anticlines extend southeast into Colorado. Paradox Valley, formed from collapse of a salt-cored anticline, contains Paradox Creek, which is a consequent creek flowing southeast into the Dolores River (west Paradox Creek) or northwest into the Dolores River, but the Dolores River flows northeastward directly across the valley and is completely indifferent to the northwest-southeast valley. Similarly, Gypsum Valley to the south contains Little Gypsum Creek (northwest) and Big Gypsum Creek (southeast), which flow as they should parallel to the axis of the valley. But the Dolores River crosses the valley at nearly right angles,

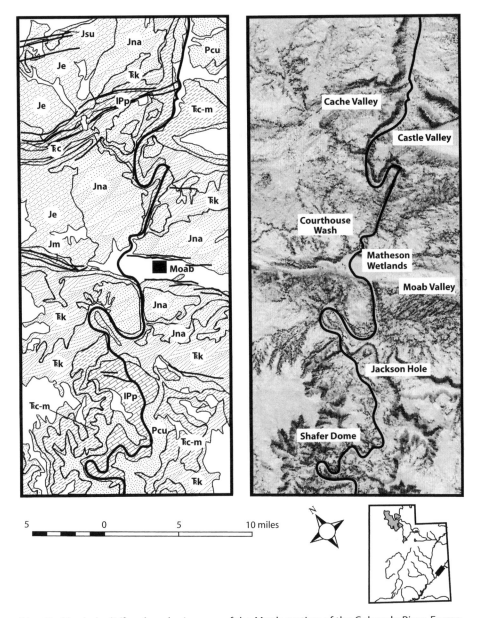

Map 21. Shaded relief and geologic maps of the Moab section of the Colorado River. Formation abbreviations are: Jsu = Summerville Formation, Je = Entrada Formation, Jna = Navajo Sandstone, Ŧk = Kayenta Formation, Ŧc-m = Chinle and Moenkopi formations, Pcu = Cutler Group, IPp = Paradox Formation.

although in this case the Dolores follows the valley northwest for 2 miles before leaving on its journey to Paradox Valley. The Dolores River is antecedent to the Dolores salt anticlines (Late Tertiary to Quaternary) (Hunt 1969).

At the Dolores River confluence with the Colorado River, the elevation is about 4,101 feet. At Dewey Bridge on Utah Highway 28, the Colorado River has barely descended. The river makes a great meander bend at Ninemile Bottom, and crosses U.S. 191 at Courthouse Wash at an elevation of 3,926 feet (Map 21).

Moab to the Needles

The Colorado River flows into Spanish (Moab) Valley, a collapsed salt anticline. At Moab, the river begins a meandering course southwest against the regional northward dip of the rocks (Map 21). The river gradient is about 1 foot per mile. The outer gorge rim is 8 to 10 miles wide, and the inner gorge is 1 to 2 miles wide. The river flows from the flat Moab Valley past the Matheson Wetlands on the south bank and the Atlas uranium tailings on the north bank. The Colorado is marshy without gravel-covered terraces, suggesting recent subsidence has locally affected the river gradient.

The Colorado River makes one bend to the south at the west wall of the valley, and exits the valley through the Portal into cliffs of Wingate, Kayenta, and Navajo Sandstone. The river then makes a big meander bend to the north around Amasa Back (capped by Navajo Sandstone) into the Moenkopi and the upper Permian rocks at Jackson Hole (Map 21). West of Moab, the Jackson Hole meander is abandoned at an elevation of 4,108 feet while river level is 3,914 feet.

The gently inclined strata are interrupted by several gentle anticlines and synclines. The river then flows into the Elephant Canyon Formation in Shafer Basin on the axis of Shafer Dome. The Colorado River wanders in and out of northeast-trending Shafer Dome, a salt-bulge anticline that lies below Dead Horse Point. The river curves right to the axis of Shafer Dome and the Cane Creek anticline with the river bed in Elephant Canyon Formation. The river makes a large gooseneck bend at Shafer Canyon. The river is indifferent to all of these structures.

The Cane Creek anticline yielded oil in the past and now yields potash from solution mining of the Paradox Member of the Hermosa Formation. The river then heads south, crossing the Lockhart anticline and Rustler Dome with just a narrow ribbon of Elephant Canyon Formation along the river. The river crosses Rustler Dome and Gibson Dome with a broad meander bend in the Pennsylvanian Honaker Trail Formation. The river begins

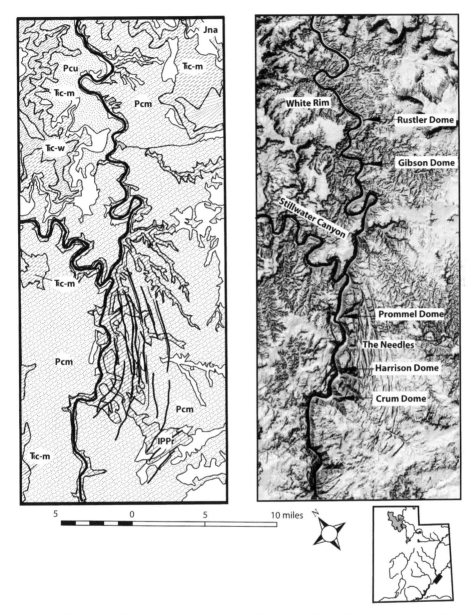

Map 22. Shaded relief and geologic maps of the Needles section of the Colorado River. Formation abbreviations are: Jna = Navajo Sandstone, T̄c-w = Chinle to Wingate formations, T̄c-m = Chinle and Moenkopi formations, Pcm =Cedar Mesa Sandstone, IPPr = Elephant Canyon Formation.

following the axis of the Meander anticline and reaches the confluence with the Green River (Map 22).

The Needles to Lake Powell

At Spanish Bottom, a dome of gypsum, the Prommel Dome appears, followed in succession by the Harrison and Crum domes. To the east a series of graben in Permian Cutler Formation are formed as rocks slide on Paradox Formation into the river on the flank of the Monument Uplift.

At an obvious diapir at the mouth of Red Lake Canyon in Spanish Bottom at the head of the Cataract rapids, sediments are punched upward by contorted gypsum and shale to form the Prommel Dome (Map 22). The Harrison Dome is in Tilted Park about 4 miles downriver where tar seeps occasionally occur. Crum Dome is at the mouth of an unnamed canyon. All three domes occur along the flank of the northeast-trending Meander anticline, a low-relief but sharp-crested anticline that extends upriver to the northeast along the river. The domes are on the axis of the valley anticline about 200 feet above the floor of the canyon and are highly eroded with all soluble salts removed. The intrusion of the salt plugs postdates the cutting of Cataract Canyon (Map 23). Each of the diapirs is structurally and stratigraphically controlled. Their location is controlled by weakness associated with fracturing along the crests of the anticlines, and the thickness of the Honaker Trail Formation is minimized by doming and topographic beveling of the domes at the floor of the canyon (Huntoon 1988). Unloading by river erosion may have caused the Meander anticline to form (Baars 2000). The typical valley anticline is in a tributary to Cataract Canyon, and sedimentary layers are arched up as canyon walls close vice-like. The process is active today.

In the Needles District, a complex of pinnacles, formed by weathering along fractures, formed at the crest and on the western flank of the Monument Uplift immediately west of the axial region (Map 22). The fracture system is enhanced by down-dip gravity sliding that forms slump blocks of "The Grabens." Canyon cutting provided the topographic relief that induced stresses that facilitate salt flowage and overburden gliding.

In the Canyonlands area, the Permian surface was stripped of Triassic cover by the end of the Miocene or beginning of the Pliocene. Erosion of the Triassic shales removed confining layers and enhanced groundwater circulation to promote salt solution. Salt flow and solution of salt in the Paradox Formation has been periodic since the salt was deposited in Pennsylvanian. The dome and graben structures described here are less than 5 million years old. Remnants of shallow Miocene-Pliocene channels eroded on the Perm-

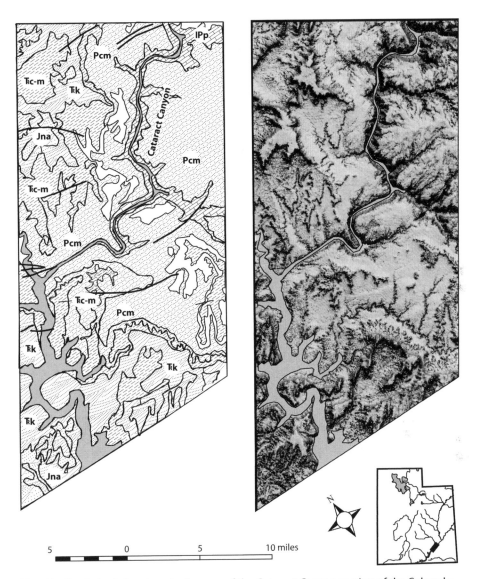

Map 23. Shaded relief and geologic maps of the Cataract Canyon section of the Colorado River. Formation abbreviations are: Jna = Navajo Sandstone, Ŧk = Kayenta Formation, Ŧc-m = Chinle and Moenkopi formations, Pcm =Cedar Mesa Sandstone, IPp = Paradox Formation.

ian surface at Cataract Canyon are preserved on the surface of horst blocks in the Needles fault zone as abandoned hanging valleys.

The Colorado River gradient is less than 1 foot per mile above and below Cataract Canyon and 8 feet per mile in Cataract Canyon (Figure 15). The increased gradient is caused by the Needles fault zone, river steepening to keep pace with Meander anticline growth, and gliding of the canyon walls inward on the Paradox salt beds. The Big Drop rapids in Cataract

Figure 15. Cataract Canyon on the Colorado River in Canyonlands National Park. Major rapids are visible from the vicinity of Standing Rocks. U.S. Geological Survey and U.S. National Park Service photograph by S. W. Lohman, 1960. The mesa and cliffs are the Permian Cedar Mesa Sandstone member of the Cutler Group. The Pennsylvanian Paradox Formation member of the Hermosa Group is exposed at river level. The Colorado River Canyon has narrowed by sliding on the Paradox Formation yielding the rapids. Much of the lower part of Cataract Canyon is beneath Lake Powell.

Canyon were formed as the canyon walls closed on river debris and the canyon floor arched upward.

According to the National Park Service, the highest recently recorded flow in Cataract Canyon was 114,000 cubic feet per second in 1984, but prehistoric flows have left driftwood piles that suggest flows of 225,000 cubic feet per second. Historically, peak flows have varied from 19,860 minimum in 1963 to the 114,000 cubic feet per second in 1984.

The last of the Paradox Formation is exposed in Gypsum Canyon. From there southward, a narrow band of Honaker Trail Formation is exposed along the river. The Cutler Formation (Permian) caps the mesa.

At Elephant Canyon the river is at 3,867 feet, and the plateau above is about 4,800 feet, representing about 1,000 feet of incision. At the Green River confluence the elevation is 3,835 feet; at Spanish Bottom, 3,832 feet; at Imperial Valley–Calf Canyon the elevation is 3,694 feet; and the water elevation of the long arm of Lake Powell at this location is at about 3,694 feet varying with year and season. The Rincon is an abandoned meander near the south shore of Lake Powell at about 3,700 feet. The Rincon river bed is at 3,868 feet, and the mesa is at 4,430 feet. U.S. 89 crosses at Glen Canyon Dam with the lake at 3,700 feet, and the river below the dam is at an elevation of 3,097 feet.

Rivers are more effective at erosion when they are high above their ultimate destination, the ocean base level, and when the gradient is steep. As base level is approached, rivers become sluggish, carry finer sediment, and deepening of their valleys ends. Rivers of the Colorado Plateau are high above sea level, and the water is fast enough to carry enormous amounts of sediment and even large boulders. The canyons are deep and getting deeper. Reservoirs such as Lake Powell or Flaming Gorge change all this by producing a temporary base level that stops canyon deepening and begins to fill canyons with sediment (Baars 1993).

Canyons are located on the easiest course for rivers to follow. Geologic structure plays a role. Rivers should flow around uplifts and into basins or at least down the dip of the rocks, but the Green and Colorado rivers ignore these simple rules. They flow south against the tilted structural grain that resulted from the northward tilt of the Colorado Plateau in early Tertiary. Rivers cut across nearly every uplift on the plateau. The Colorado River has crossed the Dry Gulch fault, the Ryan Creek fault zone, the Uncompahgre fault (buried), the Onion Creek diapir, Fisher Valley salt-cored anticline, the Cache Valley salt-cored anticline, Elephant Butte folds, Castle Valley salt-cored anticline, the Moab anticline, the Moab fault, the Monument Uplift, the Cane Creek anticline, and the Shafer Dome (as shown on geologic maps of Doelling [2001] and Huntoon, Billingsley, and Breed [1982]). The river has ignored each of these structures.

The Colorado crosses the Uncompahgre Uplift at right angles, then crosses the nose and flows along the flanks of the Monument Upwarp. Then the Colorado crosses the Kaibab Uplift at right angles in Grand Canyon. The Dolores River crosses Gypsum and Paradox Valley salt anticlines at right angles.

The rivers are superimposed on these structures. The superimposed course of these rivers was established on flat Cretaceous and Tertiary rocks that buried older structures. The rivers established their meandering course on the young rocks when the gradients were low, and deepened their canyons into young strata as uplift continued. They eventually became trapped in their own canyons and could not escape the canyons when they reached older structures. The rivers continued to meander as uplift continued, resulting in cut-off meanders. The view from Dead Horse Point shows 2,000 feet of vertical relief along the Colorado River (Figure 16).

Figure 16. Meander bends on the Colorado River as viewed from Dead Horse Point. Author photograph.

Mesozoic and Tertiary rocks greater than 1 mile thick have been removed by the rivers at a rate of 3 cubic miles per one hundred years. Dumitru, Duddy, and Green (1994) infer erosion of a considerable thickness of sediment. Apatite fission track data show that Permian strata forming the Colorado Plateau surface at the Grand Canyon underwent burial temperatures of 90 to 100 degrees Celsius in the Late Cretaceous, indicating that 9,000 to 15,000 feet of Mesozoic cover has been removed from the area. The sequence of sediments that were removed is similar to the 8,000- to 12,000-foot thickness of Mesozoic section preserved to the north in the Book Cliffs, and confirms that those strata once covered a much broader area of the Colorado Plateau. Cooling from maximum burial temperatures began about 75 million years ago. Erosion since the Laramide caused by topographic relief created by monoclines and reverse faults removed 4,000 to 15,000 feet of Mesozoic strata. Permian strata at the Waterpocket monocline in Capitol Reef National Park underwent a burial temperature of 90 degrees Celsius in the Late Cretaceous, consistent with the idea that the Permian to Upper Cretaceous section exposed in the monocline formerly extended south over the Grand Canyon region. A tentative explanation is that the Waterpocket monocline formed and eroded in Laramide time, and was then buried by Tertiary sediments that have since been removed by river erosion. Thick lower Tertiary basin sediment such as the Uinta and

San Juan basins currently cover parts of the plateau, and related cover sediments could once have extended over larger areas.

The uplift of the Colorado Plateau has altered the rates of canyon deepening by the rivers on the plateau. The rate of canyon deepening reflects the uplift rate as well as factors such as discharge, sediment load, and rock softness. Therefore, incision rates of the Colorado River are integral to understanding the development of the Colorado Plateau. The history of river incision is recorded in river terraces and gently sloping rock surfaces. Incision rates are episodic based on absolute ages of two levels of Quaternary deposits adjacent to Glen Canyon along the north flank of Navajo Mountain (Garvin et al. 2005). The minimum surface ages are determined using a variety of methods including Uranium series, cosmogenic radionuclide, and soil development (Garvin et al. 2005). Bedrock incision rates between 500,000 years and 250,000 years ago are about 1.3 feet per thousand years; from 250,000 years ago to the present, the rate is 2.3 feet per thousand years, more than double the rates in the Grand Canyon. These rates suggest that the Colorado River above Lees Ferry is out of equilibrium with the lower section of the river. Incision rates of two tributaries, Oak Creek and Bridge Creek, which flow from Navajo Mountain into Glen Canyon from the southeast, are 2 feet per thousand years over the past 100,000 years at points 5.6 miles away from the main stem of the Colorado River.

Incision of the Ancestral Colorado in the Navajo section began about 10 million years ago in response to uplift of the southern margin of the Colorado Plateau (Patton et al. 1991). Accelerated downcutting entrenched the Ancestral Colorado and Little Colorado rivers 1,100 to 1,300 feet below the Valencia Surface, a low-relief surface that predates cutting of the Grand Canyon and developed before 9.42 million years ago. Then Lake Bidahochi formed and filled with sediment in middle Miocene. Late Tertiary history of the western Grand Canyon rim gravels indicates paleodrainage was north and northeasterly across the modern Grand Canyon and fed Paleocene and Eocene lakes in eastern Utah. Western Grand Canyon was formed by headward erosion that captured the ancestral, north-flowing upper Colorado River in the vicinity of the Kaibab Uplift. The Grand Canyon formed in the last 4 million years (Patton et al. 1991).

The history of Colorado River erosion is preserved in terraces from Grand Junction, Colorado, to Moab, Utah. East of Grand Junction, the highest sloping rock surface flanking Grand Mesa projects to 1,476 feet above the river. Several pre-Bull Lake (200,000 to 130,000 years before present) flat erosional rock surfaces are 525 to 1,312 feet above the river. Two levels of Bull Lake-age erosional bedrock surfaces are 492 to 820 feet above the river level. Pinedale age erosional rock surfaces are at 66 to 197 feet above the river. Along Plateau Creek (which flows west from Grand Mesa to the Colorado River), three terrace levels capped with pre-Bull Lake fan gravel project to 196 to 492 feet above the riverbed. Near Grand Junction, Colorado River terraces 141 to 295 feet above the river are correlated with Bull Lake age glaciation. Two Pinedale (30,000 to 10,000 years before present) fill terraces occur on the Colorado in Grand Valley and the Gunnison River downriver from Delta, Colorado. In Professor Valley 21.7 miles east of Moab, the oldest terraces 82 to 105 feet above the river are 610,000 to 250,000 years old. Next oldest terraces are 39 to 72 feet above the river, and the youngest terraces 16 to 33 feet above the river are late Pleistocene. Holocene (last 10,000 years) terraces and floodplains are less than 10 feet above the river. In Negro Bill Canyon, downriver from Professor Valley, terrace gravel 328 feet above the river has reversed magnetic polarity, establishing an age of at least 750,000 years.

Pack Creek and Mill Creek head in the LaSal Mountains and flow north to the Colorado River. The oldest gravel in Pack Creek is a gravel veneer on Johnson's Up-On-Top 820 feet above the valley floor. This gravel was deposited before incision of Spanish Valley and has reversed magnetic polarity, so the age is greater than 750,000 years. At the junction of Pack Creek and Mill Creek, river terraces are 98 feet above the present river.

The Colorado River basin is arid to semiarid with limited water resources, but even so, fifty-five species and subspecies of fish are present (Tyus et al. 1982). From the Colorado-Utah border to Lake Powell, the Colorado River can be fished for channel catfish and northern pike. The many other sport fish in the river are not abundant (Prettyman 2001). The best place to seek the pike is near the Colorado border. Carp are abundant in the river from the Colorado border to Lake Powell.

The San Juan River

Spanish explorers coming from Mexico and New Mexico named the San Juan River (Van Cott 1990). Don Juan de Oñate (1552–1626) was colonial governor of New Spain (Mexico), and Don Juan Maria de Rivera led the first Spanish expedition to the San Juan area from Sante Fe, New Mexico, in 1765. The Spanish named the river San Juan Bautista (Saint John the Baptist) (Powell 1994). In Navajo mythology, the river is known as Old Age River, One-With-A-Long-Body, or One-With-A-Wide-Body (Powell 1994). Indians also called the river Pawhuska (Mad River). The Domínguez-Escalante party who crossed El Rio de San Juan in August 1776 knew the river (Warner and Chavez 1976).

The San Juan River originates on the south slopes of the San Juan Mountains in southeastern Colorado west of the continental divide near Pagosa Springs. The San Juan River flows into Navajo Reservoir, into New Mexico, then back across a corner of Colorado and into Utah. The San Juan River is 360 miles long, but only 125 miles are in Utah.

The San Juan drainage area is 25,800 square miles, but 90 percent of the flow comes from the San Juan Mountains. In flood stage, up to 75 percent of the river volume is silt and sand that is deposited in Lake Powell at the rate of 25 million tons per year (Stevenson 2000).

Not to be outdone by the Colorado and Green rivers, the San Juan River also crosses major barriers in Utah. The Monument Upwarp (Map 8), which lies directly in the path of the river, is 110 miles long in a north-south direction and 50 miles wide. The Monument Uplift is not a simple anticlinal uplift, but consists of a series of anticlines and synclines with a gentle west flank and a steep east flank expressed as Comb Ridge. The rocks are easily visible because vegetation covers less than 20 percent of the exposures. The San Juan River has carved spectacular canyons across the folds that comprise the Monument Upwarp, exposing Middle Pennsylvanian rocks in the canyon bottoms in the core of the upwarp (Baars 1973).

The San Juan River originally flowed from the south side of the San Juan Mountains and terminated in the San Juan Basin in Paleocene and Eocene time (Huber 1973). The San Juan Basin was filled with sediment by Oligocene and overflowed northwest into the Dolores River or a Dolores tributary. The Dolores River system that flowed from the eastern section of the San Juan Mountains across the upwarp to the Colorado River originally drained the Monument Upwarp (Hunt 1969). The southern drainage system into the San Juan Basin might have been captured. Due to intrusion of igneous rocks into the Ute Mountains, the Dolores was diverted north into the Colorado.

By late Miocene (23.8 to 5.3 million years ago), the San Juan had its present course across the Monument Upwarp. Earlier in its geologic history, the San Juan River bed was near base level, and the river was in the mature stage of the fluvial cycle flowing across flat-lying sedimentary beds. The river had a low gradient and numerous meanders developed. Then the area was uplifted, the gradient increased, and the river deepened its channel while still maintaining its meandering course. Layers of sediment were removed and the river was gradually superimposed on the underlying Monument Upwarp structure. The river has cut a transverse meandering canyon across four anticlines and synclines of the Monument Upwarp.

Colorado Border to Comb Ridge

Baars and Huber (1973) describe the geology along the course of the river. The San Juan River flows in a broad, alluvial valley in the Morrison Formation from the Colorado Border to Bluff, Utah (Map 24), as it approaches the eastern margin of the Monument Upwarp. Near Bluff, the San Juan River is in a wide, shallow valley between cliffs of the Bluff Sandstone Member of the Morrison Formation. The Navajo Twins form a prominent landmark overlooking the river (Figure 17). At Bluff the valley is one-half to one mile wide with flat-topped mesas bordering the valley. The mesas swing west away from the river. Gravel terraces cap the Navajo Sandstone, Carmel Formation, and Entrada Formation. A river terrace is preserved 460 feet above the river at Bluff that correlates with similar terraces that extend to Glen Canyon on the Colorado River. This terrace has been dated using cosmogenic aluminum-26 (^{26}Al) and beryllium-10 (^{10}Be). The cosmogenic age has been interpreted to represent a depositional age for the alluvial terrace of 1.36 million years. This terrace indicates a period of sediment deposition from increased Pleistocene erosion in the San Juan Mountains in Colorado, and the length of time the river required to cut down to its present level (Wolkowinsky and Granger 2004).

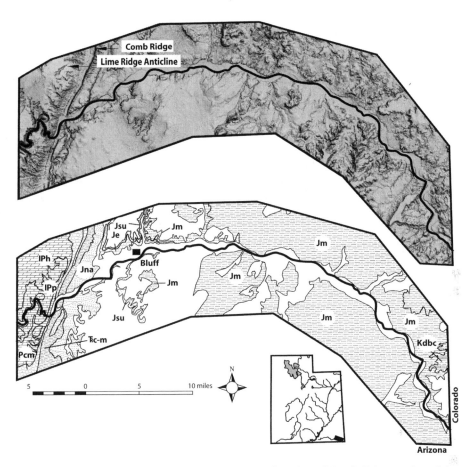

Map 24. Shaded relief and geologic maps of the Bluff, Utah, and Comb Ridge section of the San Juan River. Formation abbreviations are: Kdbc = Dakota Sandstone and Burro Canyon Formation, Jm = Morrison Formation, Jsu = Summerville Formation, Je = Entrada Formation, Jna = Navajo Sandstone, Ћc-m = Chinle and Moenkopi formations, Pcm =Cedar Mesa Sandstone, IPh = Honaker Trail Formation, IPp = Paradox Formation.

Comb Ridge to Lake Powell

Below Bluff, the river crosses into the Monument Upwarp at Comb Ridge. The gradient below Bluff is about 8 feet per mile with a few class II and class III rapids. In open stretches, bars of sand and gravel border the river and the channel width in San Juan Canyon is 150 to 300 feet, but in some places the channel narrows to 50 to 75 feet. Gravel terraces are visible up to 600 feet above river level. Boulder bars built by side canyons during flood stages, mass wasting of canyon walls, and constriction of the channel have created eighteen rapids. The average gradient is 7 feet per mile, and the discharge varies from a meager 4 cubic feet per second that barely wets the riverbed to a muddy torrent of 24,000 cubic feet per second in a single year.

Figure 17. The Navajo Twins were formerly called Punch and Judy. These two solitary knobs of the Bluff Sandporite member of the Morrison Formation are resting on columns of Summerville Formation with Entrada Sandstone at the base. The Navajo Twins overlook the community of Bluff, Utah. U.S. Geological Survey photograph taken in 1875 by W. H. Jackson. Jackson predicted the imminent collapse of the formation, but Bryan and LaRue (1927) describe the remarkable persistence.

When the river leaves the valley at Bluff, it cuts the eastern edge of the Monument Upwarp. Here, the river cuts through steeply inclined strata of the Navajo Sandstone, Kayenta Formation, Wingate Formation, Chinle Formation, Moenkopi Formation, and Cedar Mountain Formation exposed on the eastern flank of Comb Ridge (Map 24). The head of San Juan Canyon begins below Comb Ridge. The river is entrenched in a meandering course across the Lime Ridge anticline (Map 24) and tightly folded Raplee anticline (Map 25), and flows past two abandoned perched meanders. The core of the anticlines exposes the Paradox Formation, and Honaker Trail Formation is exposed in the intervening synclines. The river emerges from the Raplee anticline and crosses the Mexican Hat syncline in younger Honaker Trail Formation. From Comb Ridge to Soda Basin, the gradient is 9.4 feet per mile.

Soda Basin near the core of the Raplee anticline (Map 25) is different because of evaporite beds in the Akah Cycle of the Paradox Formation. Salt is leached from Akah rocks causing the Honaker Trail to collapse and broaden Soda Basin. Soda Basin is much like Cataract Canyon on the Colorado with Toreveh Blocks sliding toward the river. From the south edge of

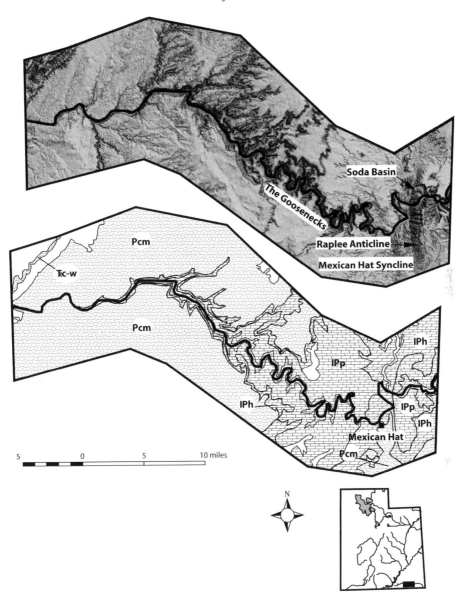

Map 25. Shaded relief and geologic maps of the Goosenecks and Mexican Hat sections of the San Juan River. Formation abbreviations are: Ʈc-w = Chinle to Wingate formations, Pcm = Cedar Mesa Sandstone, IPh = Honaker Trail Formation, IPp = Paradox Formation.

Soda Basin to Mexican Hat, the river emerges from the Raplee anticline and crosses the Mexican Hat syncline in Honaker Trail Formation.

Below Mexican Hat and the Mexican Hat syncline, the river crosses into the main portion of the Monument Upwarp. From Mexican Hat to Clay Hills Crossing (Map 26), the river is deeply entrenched into Pennsylvanian rocks in a series of entrenched meanders known as the Goosenecks.

The meandering course of the river was established long ago when the river flowed across a broad floodplain with a low gradient. Later, slow uplift caused the river to gradually entrench itself into the floodplain sediments and underlying bedrock until the river is now a prisoner between steep cliffs. The Goosenecks are probably the best known and most photographed entrenched meanders in the region.

The Honaker Trail, which provides access down the cliffs of San Juan Canyon, is located just past the Goosenecks. Above the deeply entrenched river, colorful buttes and mesas have formed in the Permian-age Cedar Mesa Sandstone Member of the Cutler group.

At the Goosenecks, the boundary between Permian rocks and Pennsylvanian rocks established by fossils is about 150 feet below the rim of the canyon. The river cuts into several of the main cycles of sedimentation in the Honaker Trail Formation including the Ismay, Desert Creek, Akah, and the top of the Barker cycles. The cyclic sedimentation is interpreted to result from advances and recessions of glaciation on the ancient continent of Gondwana. Because of the wrinkled nature of the Monument Upwarp, the river goes into and out of some of the cycles before leaving the Pennsylvanian sedimentary rocks and returning to Permian Cedar Mesa Sandstone after crossing Grand Gulch.

In the Goosenecks, the channel is 50 to 200 feet wide and the canyon depth ranges from 1,300 feet near Honaker Trail to 2,000 feet at Cedar Point. From Honaker Trail to Slick Horn Gulch, the gradient is 8.6 feet per mile. The canyon walls here lower to 1,300 feet of alternating sheer cliffs and steep slopes. From Slickhorn Gulch to Grand Gulch, the gradient is 11.6 feet per mile, then 5 feet per mile to Clay Hills Crossing where the river comes out of San Juan Canyon, now under Lake Powell (Map 26).

At Clay Hills Crossing (Map 26), the San Juan River emerges from the core of the Monument Upwarp in Permian Cedar Mesa Sandstone and enters Lake Powell. Cliffs, buttes, and mesas above Lake Powell, which occupies the San Juan River bed, are composed of Moenkopi, Chinle, Wingate, Kayenta, and Navajo strata. Lake Powell is narrow and sinuous, reflecting the identity of the San Juan River Canyon.

The youthful incision in the San Juan River system is shown by downcutting of rivers during the Pleistocene north of the San Juan River in the Abajo Mountains (Map 8) as described by Patton and colleagues (1991). The oldest Quaternary deposits in the Abajo Mountains are alluvial fan and gravel deposits that cover a smooth bedrock surface with reversed magnetic polarity. The reversed polarity establishes their age as older than 750,000 years. The deeply dissected tributary, Montezuma Creek, has cut down as much as 1,199 feet (Patton et al. 1991).

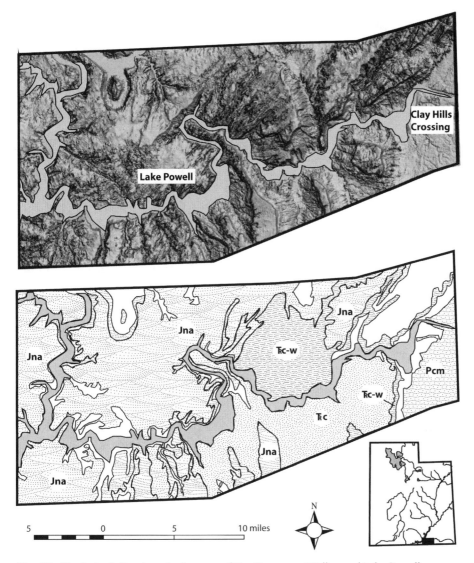

Map 26. Shaded relief and geologic maps of the Monument Valley and Lake Powell sections of the San Juan River. Formation abbreviations are: Jna = Navajo Sandstone, Ŧc = Chinle Formation, Ŧc-w = Chinle to Wingate formations, Pcm = Cedar Mesa Sandstone.

Ryden (2000) has monitored San Juan River fish. The river harbors twenty-six species and three hybrid suckers. Of these, seven are native and nineteen are introduced species. Most of the fish (99.1 percent) consist of six species: the native flannel-mouth sucker, bluehead sucker, speckled dace catfish, nonnative channel catfish, common carp, and red shiner. Rainbow trout and brown trout were introduced in upriver reaches, but of the 242,163 fish caught by Ryden (2000) only 204 were brown trout.

The Weber River

The Weber River begins in the high western Uinta Mountains at the foot of Bald Mountain (11,943 feet) and Reids Peak next to Utah Highway 150. The summit of Bald Mountain offers a view of the headwaters of the Weber, Provo, Duchesne, and Bear rivers. The Provo River begins at the foot of Bald Mountain two miles south, the Hayden Fork of the Bear River begins at the foot of Hayden Peak two miles northeast, and the Duchesne River begins at Mirror Lake one mile southeast. The Bear, Provo, and Weber rivers all discharge into Great Salt Lake but follow very different routes to get there. The Duchesne River is a tributary of the Colorado via the Green River that empties into the Pacific Ocean at the Gulf of California.

The Weber River originates in a broad, glaciated valley carved into the quartzite and shale of the Precambrian Uinta Mountain Group (Map 27). The river flows northwest through the Precambrian, Mesozoic, Pennsylvanian, and Permian sediments on the flank of the Uinta Uplift to the north-flank fault zone that is concealed by glacial deposits and Tertiary sediments. The river turns west and then south of west along the concealed south-flank fault, then flows south of west in a broad, alluvial valley in Mesozoic sediments.

The canyon narrows in the Nugget Sandstone, then broadens as the river enters Kamas Valley, a broad faulted valley filled with Quaternary sediments. The river flows west to the west side of the valley then veers north along the Keetley Volcanics contact. The river follows this contact northwest to Rockport Reservoir, which is in thrust-faulted Mesozoic sediments. The river crosses the thrust faults and flows northward through Keetley volcanic rocks and Tertiary and Cretaceous sedimentary rocks to flow into Echo Reservoir. The outlet of the reservoir is in Cretaceous Echo Formation. The river flows northwest into Cretaceous sediments, then makes a sharp bend to the south of west at the confluence with Lost Creek.

Lost Creek begins on Horse Ridge, flowing west, then south, then south-

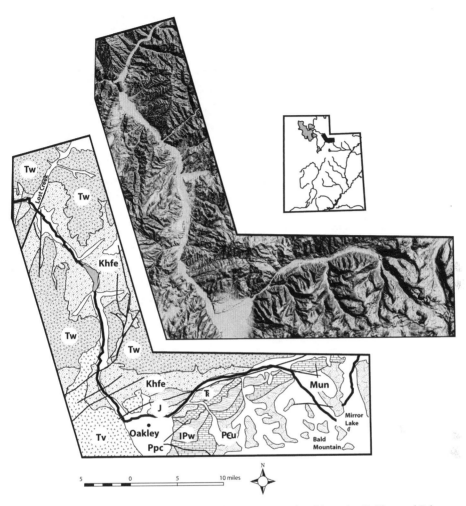

Map 27. Shaded relief and geologic maps of the upper Weber River, the Oakley and Echo Reservoir section of the Weber River. Formation abbreviations are: Tv = Volcanic rocks, Tw = Wasatch Formation, Khfe = Frontier, Henefer, and Echo Canyon conglomerate, ℞ = Triassic formations, Ppc = Park City Formation, IPw = Weber Sandstone, Mun = Mississippian formations, Pεu = Uinta Mountain Group.

east to Lost Creek Reservoir, then to the confluence with the Weber River near Devils Slide. Here the Weber River has cut a narrow canyon through steeply dipping Twin Creek Limestone in normal-sequence downsection to Cambrian sedimentary rocks, then crosses a fault into a broad north-south valley of Tertiary sediments and veers northwest in this valley. At the confluence with Cottonwood Creek, the Weber River veers due west again to enter the very narrow Devil's Gate Canyon in Precambrian Farmington Complex crystalline rocks (Map 28, Figure 18). The river emerges from this canyon into Lake Bonneville sediments and flows northwest, north,

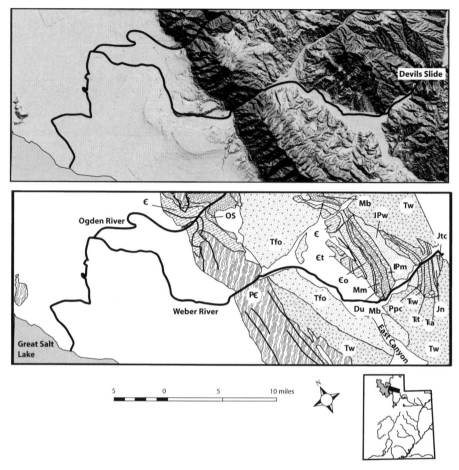

Map 28. Shaded relief and geologic maps of the Wasatch Mountains, Devils Gate and mouth of the Weber River. Formation abbreviations are: Tfo = Fowkes Conglomerate, Tw = Wasatch Formation, Jtc = Twin Creek Limestone, Jn = Nugget Sandstone, Ŧa = Ankareh Formation, Ŧt = Thaynes Limestone, Ŧw = Woodside Shale, Ppc = Park City Formation, IPw = Weber Sandstone, IPm = Morgan Formation, Mb = Brazer Dolomite, Mm = Madison Limestone, Du = Devonian formations, OS = Ordovician and Silurian formations, Є = Cambrian formations, Єo = Ophir Formation, Єt = Tintic Quartzite, PЄ = Farmington Canyon Complex.

northwest, then southwest, first to the confluence with the Ogden River and finally into Great Salt Lake.

Both Weber River and Provo River traverse parts of Heber Valley, one of the valleys on the east side of the Wasatch Mountains. The evolution of drainage on the east slope of the Wasatch and Heber Valley is a part of the Weber and Provo rivers' history. Remnants of erosion surfaces and paleo-channels show rivers on the eastern slope of the Wasatch Mountains were consequent rivers flowing eastward into the Weber River north of Heber Valley before normal faulting in Heber and Kamas Valley (Nelson and

Figure 18. Devils Gate in Weber Canyon of the Weber River. The Weber River has cut a steep, narrow canyon through the metamorphic rocks of the Farmington Complex to enter the valley of the Great Salt Lake. A railroad and an interstate highway now occupy Devils Gate. U.S. Geological Survey photograph by W. H. Jackson, 1869.

Krinsky 1982). Farther south subsequent to relative lowering of Heber Valley, the rejuvenated Provo River allowed drainage from the Strawberry area into Heber Valley to cut Daniels Canyon (Map 9). Faulting of the back valleys periodically disrupted river channels in the Weber system, and greater than 650 feet of fill accumulated in the back valleys.

The Weber was the first river to cut through the Wasatch Mountains, and of the Weber and Provo system, the Weber is the master river. Most of the drainage basin east of the Wasatch Mountains including much of the high western Uinta Mountains was developed by the Weber River before the Provo River cut through the Wasatch (Anderson 1915). The headwaters of the upper Provo River above Kamas Valley were formerly the south fork of the Weber River that united with the north fork at Peoa. Between Peoa and Rockport, the Weber channel is in nearly vertical sandstone and conglomerate, and the river flows at right angles to the strike. Erosion and channel incision were slow here and the Weber River gradient was low, so the Weber developed a sluggish, meandering course in Kamas Valley that slopes barely

200 feet from south to north end. The Provo River was slow and sluggish through Heber Valley, but the river had two forks at Hailstone. The left fork captured the Weber at the south end of Kamas Valley where the Provo was 200 feet lower than the valley level. The captured Weber channel, now a part of the Provo River, was entrenched upriver for 8 miles, and 50 percent of the headwaters of the Weber were diverted to the Provo. The entrenching is still going on, suggesting the capture was very recent (Anderson 1915).

The Weber has also gained flow from the Bear River and from Chalk Creek (Hansen 1969). The Weber captured drainage from the West Fork of the Bear in a minor diversion north of Holliday Park. Near Holliday Park, the Weber is 1,200 to 1,500 feet lower than the West Fork of the Bear River. The Weber has a steeper gradient, giving it a hydraulic advantage. Larrabee Creek eroded headward to the north and intercepted and captured the east-flowing drainage of Windy Peak, which was diverted south from the Bear to the Weber River. This minor event illustrates a dynamic evolving landscape (Sears 1924).

Soil profiles containing 23 to 75 percent carbonate are preserved on terraces 124 to 384 feet above the Weber River and on alluvial fans in Kamas Valley. These terraces and alluvial fans represent an aggradational episode related to glaciation in the Uinta Mountains more than 200,000 years ago. Similar fan remnants are found in Heber Valley.

Lower terraces 13 to 39 feet above the Weber River and uppermost Provo River grade into hummocky moraines in the river's headwaters, showing they date from the last glaciation. Heber Valley and northern Kamas Valley were covered with outwash during the last glaciation. Diversion of the upper Provo River into Heber Valley may have been aided by extensive outwash in southern Kamas Valley, and this diversion may predate the last glaciation. The major tributaries of the Weber River include Beaver Creek, Chalk Creek, Lost Creek, and East Canyon Creek; the largest tributary is the Ogden River.

The Weber River flows 70 miles from the Uinta Mountains to the confluence with the Ogden River. The river contains brown, brook, cutthroat, and rainbow trout and mountain whitefish. The lower stretch from Riverdale to Weber Canyon has some long riffles and big holes with big brown trout, but water levels are highly variable and dependent on irrigation demands. From Weber Canyon to Echo Reservoir, the river is loaded with mountain whitefish and brown trout. From Echo Reservoir upstream to Rockport Reservoir, public access is good and fishing for big brown trout is good. Above Rockport Reservoir to the headwaters, much of the river is on private property and access is difficult. Above Oakley, the river contains both brook and cutthroat trout. The upper Weber River in the Uinta Mountains and its small tributaries contain brook trout that escape from the Uinta mountain lakes (Demoux 2001; Prettyman 2001).

Chalk Creek

The headwaters of Chalk Creek (Map 9), a Weber River tributary, flow north and northwest into Wyoming, then turn south and southwest back into Utah. The north-flowing reach is separated from the Bear River by a low divide, but the southwest reach is in a deep canyon en route to the Weber. The head of Chalk Creek formerly flowed into the Bear River directly, or flowed into Yellow Creek and then into the Bear River, but was diverted by headward erosion of lower Chalk Creek and capture of the upper part of the Chalk Creek drainage. The diversion point is an elbow at the Wyoming state line. Between the diversion point and Pine Cliff, barbed tributaries reveal the reversal of the flow direction.

Chalk Creek provides excellent fishing for cutthroat trout and stocked rainbow trout, but there is little public access (Demoux 2001).

Beaver Creek

The headwaters of the present Provo River once flowed down Beaver Creek (Maps 29, 30), which is an underfit river and a Weber River tributary. A tributary of the Provo eroded northeast through Pine Valley and beheaded Beaver Creek. The saddle at the head of Beaver Creek marks the drainage divide between the Weber River and Provo River, which discharge at points 65 miles apart.

Beaver Creek contains small wild cutthroat trout and some planted rainbow trout (Demoux 2001).

East Canyon Creek

East Canyon Creek (Map 9) begins in Parleys Park and flows south toward Interstate 80 opposite to the north-flowing Silver Creek about 1 mile to the east. The divide between the two creeks is about 100 feet high, suggesting that the 180-degree turn in East Canyon Creek may be due to capture of a former Silver Creek tributary. East Canyon Creek then curves around Kimball Mountain to the west, northwest, then north in Preuss Sandstone, Kelvin Formation, Frontier Formation, Wasatch Formation, and Norwood Tuff to East Canyon Reservoir in the East Canyon Creek graben. The creek flows out of the reservoir in Preuss Sandstone, Echo Canyon Conglomerate, Wasatch Formation, and Norwood Tuff, and flows northwest to Porterville, Richville, and to the confluence with the Weber River.

Ogden River

The three forks of the Ogden River join in Pineview Reservoir near Huntsville. The North Fork begins on the back (east) side of the Wasatch Mountains in Precambrian-age Mineral Fork Tillite, an ancient glacial deposit. The river flows in an arrow straight course with no meanders through the back valley graben extension of the Cache Valley graben past Liberty and into Pineview Reservoir.

The Middle Fork begins in the Bear River Range in Cambrian-age limestone and curves into Tertiary conglomerate that covers much of the range. The fork then cuts into Precambrian Mutual Formation and then into Quaternary sediments of Ogden Valley and empties into Pineview Reservoir.

The South Fork begins in the Monte Cristo Range where the left and right forks join in faulted Paleozoic sediments that have been exposed by stripping away the overlying Tertiary conglomerate. The South Fork flows through the Tertiary Conglomerate then Cambrian sediments into the Precambrian Mutual Formation and then into the Quaternary fill of Ogden Valley and into Pineview Reservoir. From Pineview Reservoir, the Ogden River cuts through the Wasatch Mountains to join the Weber River just above Great Salt Lake (Map 28).

The Ogden River contains a large population of wild brown trout, estimated by the Department of Wildlife Resources (DWR) to be as many as 5,000 fish per mile in Ogden Canyon (Cook 2000). The Ogden River may be the most productive stream in northern Utah (Demoux 2001). The river also contains rainbow and cutthroat trout and whitefish. Whirling disease is present. Look for congregations of mountain whitefish in the deeper holes (Prettyman 2001).

The Provo River

The Provo River is a pirate and a thief. The Provo captured the upper Weber River like a spider web catches a fly. The Middle Fork and North Fork of the Provo River originate in glacial lakes in High Uinta Wilderness areas (Map 29). The Middle Fork begins at Trial Lake and the North Fork begins in a group of lakes that includes Duck Lake on the west to Wier and Marjorie Lakes on the east.

The Middle Fork flows generally south and southeast in a broad, glaciated canyon carved into Precambrian Uinta Mountain Group. The Middle Fork is joined by the North Fork at the head of Pine Valley where the river makes a 90-degree bend to the south. The combined river leaves the Precambrian rocks and cuts across south-dipping Cambrian Tintic Quartzite and Mississippian- and Pennsylvanian-age limestone. The Provo River makes another 90-degree bend to head northwest where the South Fork comes in. The bend to the northwest follows the general strike of Pennsylvanian and Mississippian limestone. The bend continues along the contact of the Keetley volcanic rocks with sedimentary rocks on the west nose of the Uinta Arch at the south end of Kamas Valley (Map 30). The River then cuts directly west across a horst of Keetley volcanic rocks between Kamas and Heber Valley graben and flows into Jordanelle Reservoir (Map 31).

Closed basin fill accumulated in the Ross Creek drainage north of Heber Valley 500,000 to 700,000 years ago. Prior to construction of the Jordanelle Dam and filling of Jordanelle Reservoir, Ross Creek flowed southward past Keetley, joined the drainage from the Ontario Number Two drain tunnel, and flowed south into the Provo River. This is all underwater today. Later headward erosion at the north end of Heber Valley captured Ross Creek basin and continued eastward toward Kamas Valley.

Below Jordanelle, the Provo River enters Heber Valley and takes a right-angle bend due south through Heber Valley (Map 31). Keetley volcanic rocks are exposed on the east bank, and Paleozoic sedimentary rocks of the Wasatch Range are exposed on the west. The Provo River flows over

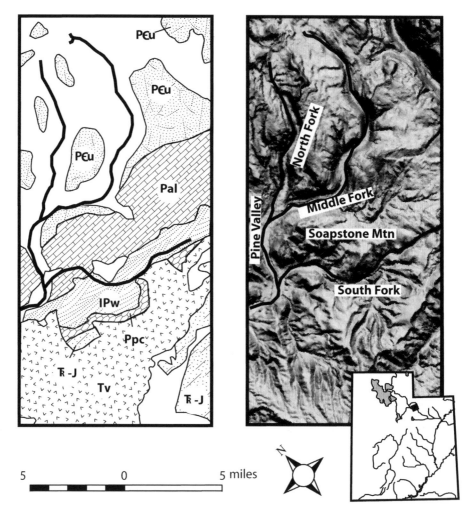

5   0   5 miles

Map 29. Shaded relief and geologic maps of the upper Provo and Pine Valley sections of the Provo River. Formation abbreviations are: Tv = Volcanic rocks, Ʀ-J = Triassic and Jurassic formations, Ppc = Park City Formation, IPw = Weber Sandstone, Pal = Paleozoic formations, PЄu = Uinta Mountain Group.

alluvial fill in Heber Valley into Deer Creek Reservoir heading southwest, then flows southwest directly across beds, across folds, across faults such as Aspen Grove faults, then flows directly across the Wasatch Mountains (Map 31). The river emerges from the Wasatch Mountains, crosses the Wasatch fault, and flows into Utah Lake (Map 32). The Wasatch fault is active today, so the river is antecedent to the fault.

The South Fork of the Provo begins at Wolf Creek Pass. Wolf Creek flows eastward from the divide to join the Duchesne River. The ridge south of the South Fork forms the drainage divide with the east-flowing West Fork of the Duchesne River. But the Provo flows west and follows the general

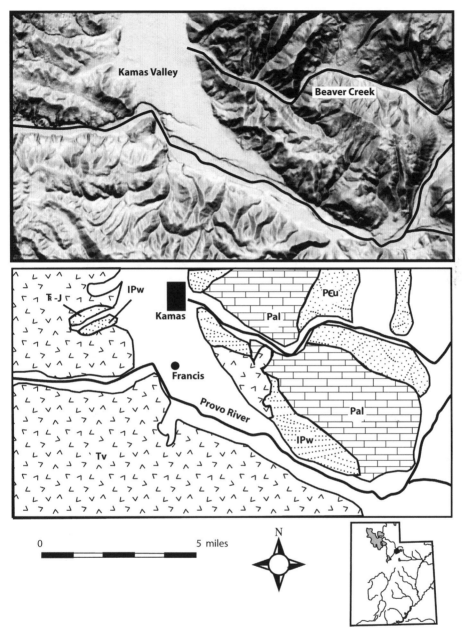

Map 30. Shaded relief and geologic maps of the Kamas Valley (Francis, Utah) section of the Provo River. Formation abbreviations are: Tv = Volcanic rocks, T̶R̶-J = Triassic and Jurassic formations, IPw = Weber Sandstone, Pal = Paleozoic formations, P∈u = Uinta Mountain Group.

Map 31. Shaded relief and geologic maps of the Jordanelle Reservoir, Heber Valley and Deer Creek Reservoir sections of the Provo River. Formation abbreviations are:
Ti = Intrusive igneous rocks,
Tv = Volcanic rocks,
T̶R̶-J = Triassic and Jurassic formations,
IPw = Weber Sandstone,
IPo = Oquirrh Formation,
Pal = Paleozoic formations.

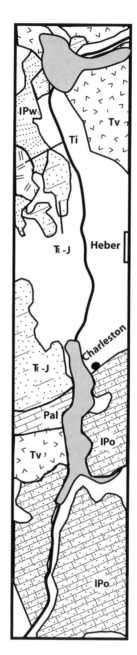

0   5 miles

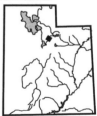

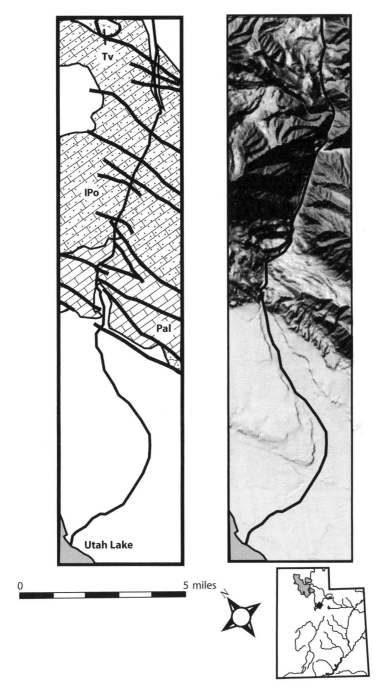

Map 32. Shaded relief and geologic maps of the Wasatch Mountains section and mouth of the Provo River. Formation abbreviations are: Tv = Volcanic rocks, IPo = Oquirrh Formation, Pal = Paleozoic formations.

strike of the Pennsylvanian Weber Sandstone northwestward to its confluence with the Middle Fork at the bottom of Pine Valley (Map 29). The river curves from northwest to west to north and back to northwest around Soapstone Mountain. The South Fork crosses the South Flank fault and is indifferent to it. The river curves to follow the curve of strike of the Weber Sandstone.

In the past the Weber River drained the entire western Uinta Mountains, but the Provo River expanded its drainage basin by capture as described by Hansen (1969). The Upper Provo flowed northward through Kamas Valley into the Weber and was part of the Weber River drainage. This portion of the Provo was diverted by capture 2 miles south of Kamas. There the present-day Provo leaves a wide, flat valley and enters a narrow canyon to Heber Valley (now a part of Jordanelle Reservoir). The capture here could have been aided by the clogging of Kamas Valley with glacial outwash. The headwaters of the Provo, once a part of the Weber River drainage, flowed down Beaver Creek, which is an underfit river and a Weber River tributary. The Provo eroded northeast through Pine Valley, beheaded Beaver Creek, and captured half of the Weber River flow.

The 50–mile stretch of Provo River from its headwaters to Utah Lake provides high-quality fishing and is one of Utah's most important and most heavily utilized fisheries. The lower stretch of river from Utah Lake to the mouth of Provo Canyon contains walleye, perch, white bass, catfish, and carp, warm-water fish that migrate up from Utah Lake. A few large brown trout are also present. From Provo Canyon to the Deer Creek Reservoir, the scenery is gorgeous, and the water and fishing are excellent for brown trout, some rainbow trout, and a few cutthroat trout—all mostly wild fish—especially in the 6-mile section below Deer Creek Reservoir. The river can be fished year round. From Deer Creek to Jordanelle, known as the Middle Provo and a favorite of fly fishers, the river contains about 3,000 fish per mile, mostly brown trout with some rainbow trout. From Jordanelle to the headwaters, the Provo contains small brown trout and cutthroat trout, as well as spectacular scenery. From Rock Cliffs State Park to Woodland and north to the Mirror Lake Highway (State Route 150), access is restricted by private property. Access to the upper Provo along the highway is good, and brown trout, cutthroat trout, whitefish, and stocked albino trout can be caught (Demoux 2001; Prettyman 2001).

The Bear River

There are three anomalous canyons that are eroded transversely through the Wasatch Mountains: the Provo, Weber, and Bear rivers. The Ogden River and Box Elder Creek drain faulted back valleys of the Wasatch and cross the Wasatch front. During the Late Cretaceous to Early Tertiary, the Great Basin was higher than the present site of the Wasatch Mountains and the Uinta Basin. Drainage from this topographic high was eastward first to the Cretaceous sea (Map 6) and later to the Tertiary lakes (Map 7). Beginning in the Oligocene, the western United States was uplifted, but the Great Basin had begun to form so the eastward-flowing rivers began to pond to form permanent lakes. Then in middle to late Tertiary the rivers reversed their flow directions to the west as the Wasatch Mountains were uplifted. The Provo, Weber, and Bear rivers are superimposed across the rising Wasatch Mountains (Hunt 1987). As the Wasatch continued to rise, the three rivers eroded antecedent canyons across the rising Wasatch, a process that continues today.

The Bear River is the largest river draining into the Great Basin, and the longest river in the Western Hemisphere that does not flow into an ocean. The longest river entirely within the Great Basin is the Humbolt River, which flows directly across the structural grain of the Great Basin and empties into Carson Sink. The Bear River provides about 40 percent of the inflow into Great Salt Lake. The river loops about 300 miles across Utah, Wyoming, and Idaho, ending in Great Salt Lake a mere 85 miles from its origin. The river flows northward from the high Uinta Mountains to Soda Point in Idaho, then through Oneida Narrows southward, and returns to Great Salt Lake in Utah like the prodigal son after wandering through parts of Wyoming and Idaho (Map 33).

Three forks of the Bear River—the East Fork, Stillwater Fork, and Hayden Fork—arise in the high, western Uinta Mountains in Precambrian rocks. The westernmost fork, Hayden, is a scant one mile east of the

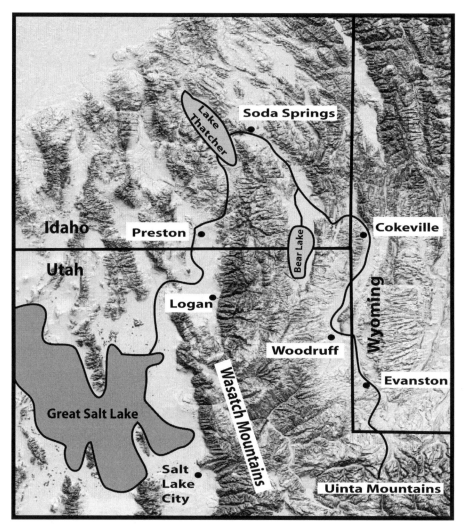

Map 33. The course of the Bear River in Utah, Wyoming, and Idaho showing the location of ancient Lake Thatcher.

headwaters of the Weber River and a few miles northeast of the Provo River headwaters. The Bear River occupies broad, glaciated valleys, but within 5 or 6 miles it crosses the concealed north flank thrust fault zone. High canyon walls become Wasatch Conglomerate and the riverbed is glacial till of the last Pinedale age glaciation. The Bear River leaves Utah flowing north into Wyoming and crosses Wyoming flowing on Tertiary-age lake sediments that conceal underlying thrust faults.

The Bear River crosses from Wyoming into Idaho near U.S. Highway 30, makes a great looping bend to the southwest, then cuts back to the northwest into Bear Lake Valley about 6 miles north of Bear Lake and the complicated confluence with Bear Lake Outlet (Map 33). A huge diversion here sends three-fourths of the Bear River flow into Bear Lake. The Bear River flows northwestward from Bear Lake valley paralleling U.S. 30 to Soda Springs, Idaho, with the Bear River Range on the west. At Soda Springs the River turns due west around the north end of the Bear River Range and flows south through Black Canyon and Gentile Valley, rather than following its former course westward to the Portneuf River and into the Snake River at Pocatello.

A rift volcano near the intersection of State Highway 34 and U.S. 30 near Soda Springs diverted the Bear River here. The valley floor was raised, and lava flowed 40 miles west to Pocatello followed by a second lava flow. Lava flows turned the Bear River southward into the Bonneville Basin. These lava flows can be seen at Grace, McCammon, Inkom, and Pocatello. The diversion by the lava dam occurred more than 34,000 years ago and may partially account for the overflow of Lake Bonneville during the Pinedale glaciation. However, the lake did not overflow during the stronger Bull Lake glaciation, which preceded the Bear River diversion.

The river continues south through Mound Valley, the site of ancient Lake Thatcher. The Thatcher Basin includes Mound, Gentile, Gem, and Portneuf valleys. The Bear River Range bounds Mound and Gentile valleys on the east, and the Portneuf Range bounds them on the west. The elevation of the highest terrace of ancient Lake Thatcher is 5,446 feet and the lower terrace is 5,102 feet. This terrace formed after Oneida Narrows gorge, 900 feet deep and 300 feet wide, was cut connecting Thatcher Basin with the Bonneville Basin (Hochberg 1996). Lake Thatcher probably existed in the Miocene and later contained the northeastern embayment of Lake Bonneville. Molluscs of the Thatcher Basin show Lake Thatcher was tributary to the Snake River in early Pleistocene (Taylor and Bright 1987).

The Thatcher Basin contained three lake episodes. During the oldest episode from 1 million years to 500,000 years ago, the lake level was low; then there is a period of no record from 500,000 to 110,000 years ago, followed by two lake high stands. The first high stand lasted for 30,000 years from 110,000 to about 80,000 years ago when the lake level was at 5,446 feet. The Bear River may have entered the Bonneville Basin at this time as water overflowed to the south. Then the second, Lake Thatcher II, began at about 40,000 years ago and receded after the divide at Oneida Narrows was cut, allowing an arm of Lake Bonneville into Thatcher Basin about 20,000 years ago. Thatcher Basin is separated from Cache Valley by a low divide (5,480 feet elevation) at the south end of Gem Valley. This divide is developed on an ancient sloping rock surface at the base of the slope that is 490 feet above the floor of Gem Valley. Here the highest stand of Lake Thatcher was 330 feet above the Bonneville shoreline in Cache Valley. Sediment deposited in Lake Thatcher includes the Main Canyon Formation that was deposited over several lake cycles from 2 million years ago to 50,000 years ago. A Pleistocene basaltic volcanic field covers northern Gem Valley and fills the former course of the Bear River when it was a major regional drainage to the Snake River.

The timing of the rise of Lake Thatcher, the southward diversion of the Bear River, and spillover to the Bonneville Basin are still unclear. Oneida Narrows may first have been cut in the Late Miocene, backfilled with sediment and then excavated again at about 50,000 years ago. On the basis of amino-acid age of a single shell, the oldest known inflow of the Bear into Thatcher is about 140,000 years ago, but could have been earlier because the sediment record is incomplete. This dating is consistent with K-Ar dates of a basalt flow along the Portneuf Gorge near Lava Hot Springs. The evidence suggests that diversion of the Bear into Thatcher was not a single event. There were two or more diversions: an earlier diversion at 140,000 years ago that may or may not have spilled over into the Bonneville Basin, and a second diversion at 50,000 years ago that did spill over into the Bonneville Basin. By 20,000 years ago the incision of Oneida Narrows was complete and Lake Bonneville backed up to the Thatcher Basin (Link, Kaufman, and Thackray 1999). Dave Rogers, Idaho State University at Pocatello, has an Ar-Ar age of 460,000 years on Portneuf basalt.

From the site of Lake Thatcher, the Bear River flows southward through the gorge of Oneida Narrows. The Bear flows south past Preston, Idaho, and back into Utah (Map 34). Then the Little Bear River joins the Bear River at Cutler Reservoir, and the Bear River heads south paralleling the Malad River, which it finally joins in Bear River Valley. The Bear River has cut a gorge (the Gate) 1,000 feet deep through the ridge that connects the Wellsville Mountains with the Clarkston Mountains (Map 34). The Bear crosses the ridge of resistant Paleozoic sediments where the ridge is the highest. The river course dates from when Cache Valley was continuous with Great Salt Lake Valley (Hunt 1987). The course is superimposed and became antecedent when Tertiary fill in Cache Valley was raised above Salt Lake Valley and the present ridge was upfaulted. The Bear then flows into Bear River Bay of Great Salt Lake. There the Bear River supports wetlands and 250 species of migrating birds.

The forks of the Bear River, East Fork, Stillwater Fork, Hayden Fork, and West Fork flowing north from the north slope of the Uinta Mountains contain stocked rainbow, cutthroat, and brook trout. The largest fish are

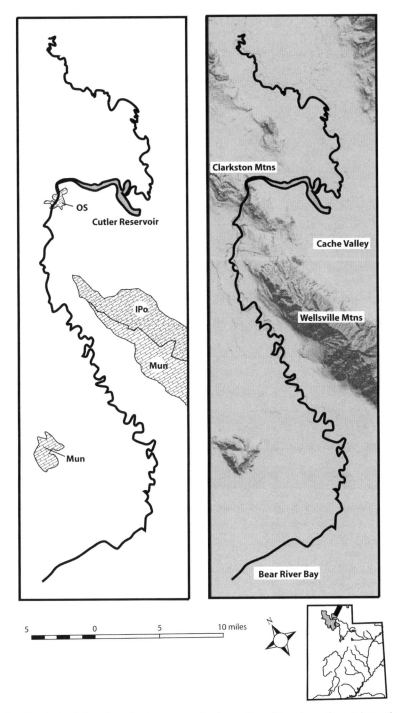

Map 34. Shaded relief and geologic maps of the lower Bear River from the Utah Border to Great Salt Lake. Formation abbreviations are: IPo = Oquirrh Formation, Mun = Mississippian formations, OS = Ordovician and Silurian formations.

found in beaver ponds (Demoux 2001; Cook 2000). After flowing through parts of Wyoming and Idaho, the Bear River reenters Utah. From the Utah line south to Great Salt Lake, the Bear River contains channel catfish, bullhead catfish, walleye, crappie, and large-mouth bass. (Prettyman 2001).

Little Bear River

The East Fork of the Little Bear River begins near the crest of the Monte Cristo Range and flows east, then crosses through the Bear River Range to Porcupine Reservoir. The Bear River Range includes the Logan Peak syncline with Pennsylvanian- to Permian-age sedimentary rocks along the axis. Many north-south normal faults affect the lower Paleozoic rocks here. The river flows east to Avon in Cache Valley and the confluence with the South Fork.

The South Fork begins on the divide between Bear River Range and the Wasatch Range near James Peak and Powder Mountain Ski Resort and flows north into Cache Valley to the confluence with the East Fork. The Little Bear then flows north to Hyrum Reservoir and northwest out of Hyrum Reservoir, and meanders into Cutler Reservoir and the confluence with the Bear River there.

The upper Little Bear River is a classic fly-fishing river for rainbow, cutthroat, and brown trout (Prettyman 2001).

Logan River

The Logan River arises in Idaho about 2 miles north of the Utah border and flows south with the Bear River Range on the west and the Wasatch Range on the east. Geology along the Utah portion of the Logan River is shown on the geologic map by Dover (1995). The Logan River enters Utah from Idaho between the Bear River Range and the Wasatch Range that separates Logan Canyon from Bear Lake on the east. The river flows south between steep mountain slopes composed of Cambrian, Ordovician, Silurian, and Devonian limestone, dolomite, and quartzite. Glacial moraine deposits and a patch of Tertiary-age Wasatch Formation conglomerate cover the high mountain slopes to the west. The river continues southwest and west to cross the Bear River Range in a steep-walled canyon carved into Ordovician, Silurian, Devonian, and Mississippian limestone and dolomite. At the west edge of the Bear River Range, the river crosses the Providence Canyon thrust fault system and the range-bounding Logan normal fault. The Logan River, the Little Bear, and the Bear River flow into Cutler Reservoir.

Blacksmith Fork

Blacksmith Fork begins at the crest of the Monte Cristo Range, which is faulted, Lower Paleozoic sedimentary rocks. Here the Wasatch Conglomerate that formerly covered the Paleozoic rocks has been stripped away. The river flows straight west and up the dip of the Cambrian-age rocks to the Ant Valley graben, a down-faulted block of Paleozoic sedimentary rocks that comprise the Ant Valley anticline. The river flows north to Hardware Ranch, then west to cross the Bear River Range in Blacksmith Fork Canyon, a steep-walled canyon in Cambrian, Ordovician, Silurian, and Devonian sedimentary rocks. The river crosses the Logan fault into the Cache Valley graben. The Providence thrust fault is buried here but exposed to the north. The river meanders over Quaternary Lake Bonneville deposits in Cache Valley northwest to the confluence with the Little Bear River.

There are two possible explanations for the Logan River, Little Bear, and Blacksmith Fork rivers' crossing of the Bear River Range. Either the crossing is superimposed as the rivers remove the Tertiary-age Wasatch Conglomerate or the crossing is antecedent as the range is uplifted along the Logan and Cache Valley faults.

High Uinta Rivers

Rivers flow from the High Uinta Mountains north into the Green River Basin and south into the Uinta Basin. These rivers are popular water for fishermen and avenues for hikers and backpackers who enjoy the scenery and solitude of the High Uinta Mountains.

Physical conditions for the rivers are different on the north and south slopes of the Uinta Mountains. Bedrock geology, river gradients, and extent and style of Pleistocene glaciation are different on the north and south slopes, but hydrometeorological conditions are similar (Carson 2005). North-slope rivers begin in a limited number of glacial cirques, but cirques on the south slope coalesce forming large, open parklands in the headwaters of rivers. The headwater basins on the south slope contain many lakes, perennial wet meadows, and straight channels. The east fork of Smiths Fork on the north-slope headwater basin is 53 square miles in comparison with the Uinta River headwater basin on the south slope with an area of 163 square miles, more than three times the north-slope basin. North-slope valley gradients decrease and widths increase, resulting in meandering or braided rivers that flow over broad alluvial floodplains, glacial outwash, and low gradient meadows into the Green River Basin. South-slope rivers flow through broad, glacially scoured parkland. They flow into the Uinta Basin through deeply scoured canyons.

Most rivers that drain the subalpine Uintas on both north and south slopes are part of the Green River system. Rivers originate from springs in talus slopes and from cirque lakes that feed into main river channels. Where resistant bedrock channels confine the valley walls, the rivers consist of cobbles and boulders. In less resistant bedrock the valleys widen into meadows floored by sand and silt with gradients less than 2.5 feet per mile.

The anticlinal structure of the Uinta Mountains cored by relatively resistant quartzite with Paleozoic Limestones on the north and south flanks leads to some unique hydrology as described by Spangler (2005). Rivers originating within the core of the mountains may lose water into carbonate rocks through karst features and blind valleys in places where bedrock consists of limestone or dolomite that is soluble in water. Acidic groundwater in such areas dissolves the bedrock producing greater avenues for groundwater flow. The result is sinkholes and blind valleys as surface water seeps into the underlying cavities in the rock. This water then discharges from large springs. Karst features include sinking and losing rivers, caves, sinkholes, pits, and large springs. Particularly well-developed karst features are present in Soapstone Basin, Blind River, and Wolf Plateau. Blind valleys have formed at the entrance to Big Brush Creek and Little Brush Creek caves. Big Spring in Sheep Creek is connected to Lost Creek 14.4 miles due west.

The course of rivers flowing from the High Uinta Mountains has been profoundly affected by glaciation. The Uintas have been glaciated several times. These glacial cycles are named for prominent localities where glacial deposits have been studied and mapped. Glacial features in the Uinta Mountains are similar to glaciation in the Wind River Mountains where the young glaciation is named Pinedale after deposits in the vicinity of Pinedale, Wyoming, and the older glaciation is called Bull Lake after deposits in the vicinity of Bull Lake on the east flank of the Wind River Mountains. Bull Lake glaciation lasted from approximately 200,000 to 130,000 years ago, and Pinedale glaciation, the last major ice age in the Rocky Mountains, lasted from approximately 30,000 to 10,000 years ago. Initial Cenozoic exposure ages date the last glacial maximum in the southern Uintas at 15,000 to 19,000 years ago (Laabs et al. 2003). During the peak of the younger glaciation, termed Smiths Fork glaciation in the Uinta Mountains, there were nineteen valley glaciers occupying an area of 363 square miles on the north flank of the Uintas. During the earlier Blacks Fork glaciation, the glaciated area was about 425 square miles (Munroe 2005). Smiths Fork-age glaciers in Ashley, Carter, and Sheep creeks at the eastern end of the Uinta Mountains experienced extensive ice stagnation. At the western end, ice did not stagnate but retreated actively after formation of terminal moraines (Munroe 2005). On the south flank of the Uintas, glaciers on the southwestern and southeastern valleys were confined to deep canyons during the last glacial maximum, but south-central drainage basin glaciers extended beyond the mountain front. The south slope was dominated by six larger glaciers with an area of more than 58 square miles (North Fork of the Duchesne, Rock Creek, Lake Fork, Yellowstone, Uinta River, Whiterock [Laabs and Carson 2005]). In comparison, there were nineteen glaciers on the north slope. The maximum ice thickness was about 1,600 feet. Seven smaller valley glaciers occupied 1.4 to 31 square miles. Moraines below Lake Fork, Yellowstone, and Uinta canyons show evidence of multiple Pleistocene advances. The youngest advance is Smiths Fork (24,000 to 12,000 years ago) and Blacks Fork (186,000 to 128,000 years ago) glaciations. An extensive lateral moraine beyond the mouth of the Yellowstone and moraines in Lake Fork and Uinta River indicate an earlier Altona glaciation (659,000 to 620,000 years ago). Most sediment on the valley floors was deposited during the last deglaciation 17,600 to 12,000 years ago.

Blacks Fork was named for Daniel Black, a member of the Ashley party. Uncle Jack Robinson built a cabin on this creek that was the earliest permanent settlement in the region (Urbanek 1988). Blacks Fork River flows northward from the crest of the Uinta Mountains into Wyoming where it flows into the Meeks Cabin Reservoir, then heads north-northeast with some canals removing water. The river follows a fairly straight course to Rollins Reservoir, then flows more easterly, crosses Interstate-80, and begins to meander. Hams Fork joins Blacks Fork at Granger. Blacks Fork continues a meandering course more to the east, then at Westvaco heads southeast and at Bryan heads due south into some large meander bends past Massacre Hill on the west bank and flows into Flaming Gorge Reservoir. Were it not for the reservoir, Blacks Fork would have joined the Green River.

The Uinta River begins in a high glacial cirque at the foot of Gilbert Peak (13,442 feet elevation), Mount Powell, and Kings Peak (13,528). The river begins as Gilbert Creek, flowing southeast over glacial moraine that covers the Precambrian Uinta Mountain Group rocks, then flows into Quaternary gravels covering Duchesne River Formation and Precambrian and Cambrian to Mississippian sediments of the south flank of the Uinta Mountains. The river flows over alluvium and gravel surfaces that cover the Duchesne River Formation to its confluence with the Duchesne River.

Glacial outwash deposits extend south from where Lake Fork, Yellowstone, Uinta, and Whiterocks drainages emerge from the Uinta Mountains. There are six separate river terraces that represent episodes of downcutting of these rivers (Reheis et al. 1991).

William Ashley named Henrys Fork in 1823 for his business partner and good friend, Major Andrew Henry (Urbanek 1988). Henrys Fork drains about 540 square miles starting in a compound cirque beneath Gilbert Peak. Kings Peak, the highest peak in Utah, lies a short distance to the south. Henrys Fork flows northeast from the high Uintas into Wyoming, flowing northeast from the Utah line and curves east just north of Lonetree, then a bit south of east collecting water from north-flowing tributaries. North of Mcinnon the river flows south of east into Flaming Gorge east of Manila, Utah.

Upper Henrys Fork dissects glacial till, has a meandering channel, and flows north through lodgepole and aspen. Farther from the foothills, the landscape is open sagebrush steppe. This unglaciated section begins about

3.7 miles south of the Utah-Wyoming border. Near Lonetree the river abruptly turns east. Then 11.2 miles down from Lonetree, the valley narrows to 1,300 feet with wide meanders on a broad grassy plain between bedrock valley walls. Henrys Fork is on bedrock for short distances between Lonetree and Burnt Fork, but mostly the river flows over alluvium. The bedrock downriver from Burnt Fork changes from Browns Park Formation to Green River Formation.

From Pleistocene age moraines down to Flaming Gorge Reservoir, the river history is recorded in nine distinct main river gravels, six mountain foot gravels, and landslide deposits (Counts and Pederson 2005). The older, higher remnant gravels cannot be directly linked to glacial units. Younger gravels can be traced from glacial till through outwash plains to river valley gravels with terraces on them. The gravel consists of fragments of the Uinta Mountain Group from the core of the range and some Paleozoic limestone from the Uinta flank. Near the moraines the gravels are thicker but quickly thin downriver and lie on planar bedrock terraces that converge downriver. The bedrock terraces and gravels have not yet been dated (Counts and Pederson 2005).

Smiths Fork was named for Jedediah S. Smith, who drew maps for Ashley and was the first white man to explore the river (Urbanek 1988). Smiths Fork flows north from the Utah border past Tipperary Bench and Poverty Flat, then curves northeast, then east, then north again and joins Blacks Fork just north of I-80.

The Duchesne River

The North Fork of the Duchesne River begins in glacial till resting on Precambrian Uinta Mountain Group rocks in the high western Uinta Mountains near Mirror Lake (west fork) and Jordan Lake (east fork) in the shadow of Bald Mountain and Reids Peak (Map 35). The river flows with a steep gradient and no meanders south down a steep, U-shaped glaciated canyon cut in the Precambrian Uinta Mountain Group past Hades Canyon, Lightening Ridge, and Rhoades Canyon, then joins the West Fork. The North Fork of the Duchesne crosses the Hoyt Canyon fault and at the confluence with Iron Mine Creek crosses the South Flank normal fault. The river then crosses in succession the south-dipping sedimentary sequence in a V, pointing downriver from Cambrian to Jurassic and into younger sediments that cover the older rocks (Map 36).

The West Fork of the Duchesne River begins in the Wasatch Mountains east of Heber City, Utah, and flows due east with no meanders to the confluence with the North Fork. At the confluence, the river heads southeast in Duchesne River Formation and parallels the east-west-striking Duchesne fault zone. The river cuts into Tertiary lake sediments and Quaternary gravels of the Uinta Basin and skirts the Starvation Reservoir. The river begins to meander as the gradient lessens. The river heads due south at Starvation to the town of Duchesne, then due east to Myton (Map 37). The river continues more or less east with more meanders. The river then turns southeast in young gravels, alluvium and Uinta Formation, fluvial gravels, and lake deposits to the confluence with the Green River at Ouray, Utah. The White River joins the Green River from the east (Map 38).

Above the town of Duchesne, Utah, the Duchesne river drainage is 660 square miles, and there are 80 miles of fishable water for rainbow, cutthroat, and brown trout (Cook 2000). Brook trout dominate the upper reaches of the river, but rainbow trout are stocked near campgrounds (Prettyman 2001). The north fork of the Duchesne near Mirror Lake in the Uinta Mountains is good fishing for rainbow, cutthroat, and brook trout.

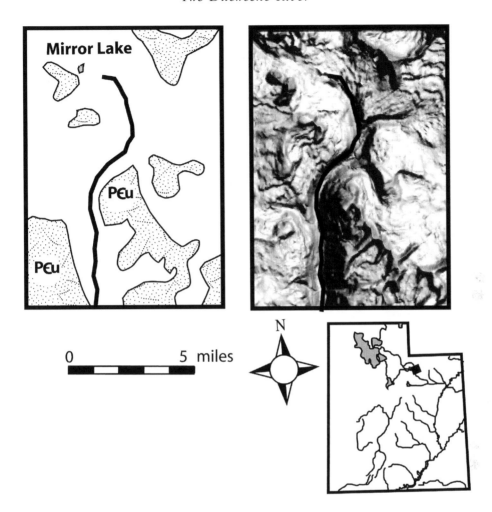

Map 35. Shaded relief and geologic maps of the upper Duchesne River. Peu = Uinta Mountain Group.

The West Fork contains some wild cutthroat trout. Below Hanna, much of the land is private or on Indian tribal lands, but with permission there is good fishing for experts (Demoux 2001).

Strawberry River

The Strawberry River begins near Currant Creek Peak and Heber Mountain on the east side of the divide with the West Fork of the Duchesne River. The river follows three strands of the Charleston thrust fault where Pennsylvanian and Permian sediments are thrust onto Mesozoic-age Woodside, Ankareh, and Thaynes formations. The river continues into the Tertiary Duchesne River Formation and into Strawberry Reservoir. The river flows

Map 36. Shaded relief and
geologic maps of the North
Fork of the Duchesne River.
Formation abbreviations are:
Tdr = Duchesne River formation,
Tu = Uinta Formation,
Tg = Green River Formation,
Km = Mancos Shale,
J = Jurassic formations,
Ҧ = Triassic formations,
Ppc = Park City Formation,
IPw = Weber Sandstone,
IPm = Morgan Formation,
Mun = Mississippian formations,
Єt = Tintic Quartzite,
PЄu = Uinta Mountain Group.

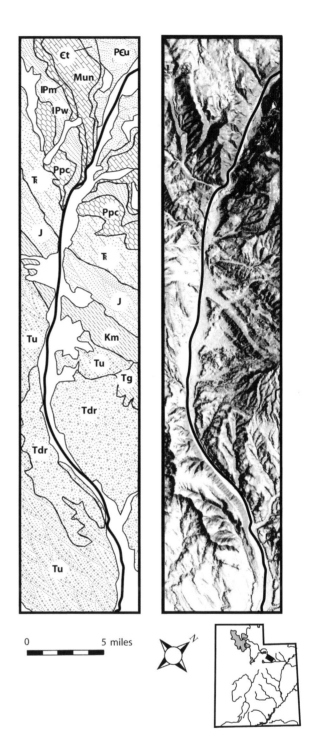

0   5 miles

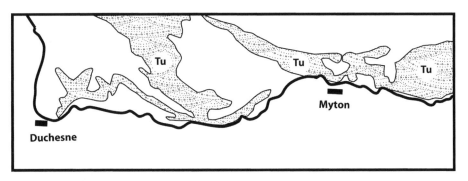

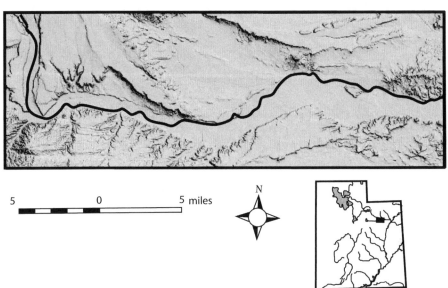

Map 37. Shaded relief and geologic maps of the Duchesne to Myton section of the Duchesne River. Formation abbreviation is: Tu = Uinta Formation.

out of the reservoir in Duchesne River Formation into Green River Formation, then into Starvation Reservoir. Strawberry River joins the Duchesne a few miles below Starvation Reservoir.

The Strawberry River above Strawberry Reservoir contains small cutthroat trout, but the river is closed to fishing from mid-May to July and from September 1 to the second Saturday in October to protect spawning cutthroat and Kokanee salmon. This reach is catch-and-release with artificial lures only. Strawberry River below Strawberry Reservoir also shares the artificial-lures-only regulation. It is primarily a brown trout fishery (Demoux 2001; Prettyman 2001).

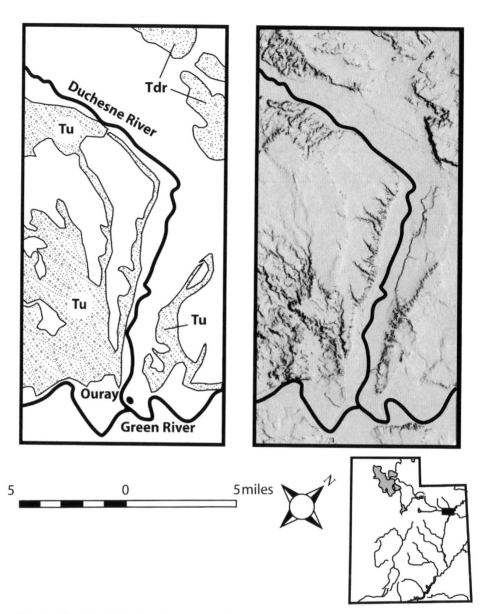

Map 38. Shaded relief and geologic maps of the Ouray section of the Duchesne River and its confluence with the Green River. Formation abbreviations are: Tdr = Duchesne River formation, Tu = Uinta Formation.

The Sevier River

The Sevier River displays much of the variability to be expected in an evolving landscape in a region of active mountain building. It flows first north, then south; sometimes it is superimposed, sometimes antecedent, sometimes diverted by piracy; and sometimes misidentified. The Sevier River is the longest river that is entirely within Utah. The name is a corruption of Rio Severo, the name used by the Spanish. It means severe and violent (Van-Cott 1990); or the name is derived from a Paiute Indian word *Seve'uu*. The name used by Domínguez and Escalante was El Rio de Santa Isabel, though their mapmaker Miera showed it on his map as a continuation of the Rio Buenaventura (Green River) (Alter 1941). For a short time it carried the name Nicollet, and Jedediah Smith named it the Ashley River.

The Sevier River Basin is a closed basin that consists of 10,522 square miles of high plateaus, narrow valleys, and broad deserts. The basin, 12.5 percent of the area of the state, is 178 miles long (north to south) and 123 miles wide. The Colorado River Basin bounds the Sevier River Basin on the east, with the Beaver River Basin on the west and the Great Salt Lake Basin on the north (Utah Department of Natural Resources 1991).

The history of the Sevier River includes episodes of capture, superposition, and antecedence as a result of the erosional history on the High Plateaus portion of the Colorado Plateau. The High Plateaus, including the Sevier Plateau, are eastward-tilted fault blocks of Cretaceous sedimentary rocks and Cenozoic sedimentary and volcanic rocks (Rowley 1968). Volcanic activity began during closing stages of deposition of the Claron Formation. Later, a laccolith was emplaced in the Claron Formation. The southern Sevier Plateau has 5,000 feet of topographic relief above valley floors due to block faulting mainly on the north-south Sevier fault. Rivers draining the plateau are in part consequent rivers that flow along structural valleys and down the dip of tilted fault blocks. Subsequent drainage has formed by headward erosion and lengthening of tributary canyons.

The Sevier River begins at a drainage divide on U.S. Highway 89 at

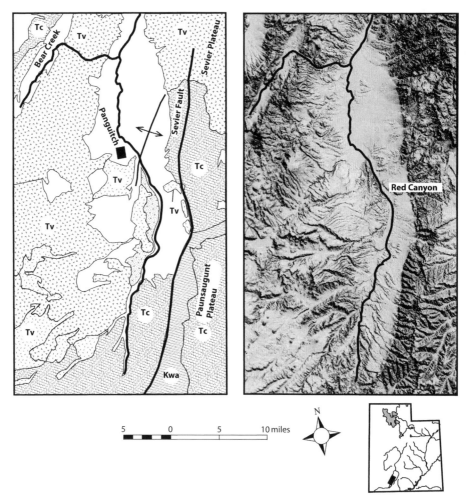

Map 39. Shaded relief and geologic maps of the upper Sevier River and the Panguitch section of the Sevier River. Formation abbreviations are: Tv = Volcanic rocks, Tc = Claron Formation, Kwa = Wahweap Sandstone.

Long Valley Junction where Utah Highway 14 goes east to Navajo Lake (Map 39). At the drainage divide (elevation near 7,500 feet), the Sevier River flows north and the East Fork of the Virgin River flows south. The Virgin River is a tributary of the Colorado River that discharges into the Gulf of California. The Sevier River ultimately discharges into the Great Basin at Sevier Lake. U.S. 89 parallels the Sevier River northward toward Panguitch, Utah. The landscape rises to the Paunsaugunt Plateau about 6 miles to the east, while the Markagunt Plateau is on the west. The Sevier fault forms the boundary between the footwall Paunsaugunt Plateau and the hanging wall river valley. High points on the plateau are above 9,500 feet. The plateau is capped with Claron Formation, the colorful lake sediment displayed in

Bryce Canyon and Cedar Breaks that comprise the pink cliffs of the Grand Staircase.

From its beginning in lake sediments of the Claron Formation, the Sevier River flows northward 6 miles to contact with basalts and other volcanic rocks and into the alluvial-filled valley of Hatch and Panguitch.

Harbor (1998) describes an episode of antecedence where the Sevier River crosses a growing anticlinal fold. The Sevier River near Panguitch crosses a narrow, northeast-trending, growing anticline and broad syncline formed in response to oblique slip on the Sevier fault. Quaternary fans or eroded rock surfaces from the footwall of the Sevier fault usually flank the river, but here the Sevier River cuts through older valley fill in the Pliocene to Quaternary-age Sevier River Formation (Anderson and Rowley 1975). In the river valley, late Quaternary alluvial fans and bedrock surfaces are being elevated at the anticline and submerged at the adjacent syncline. The anticline has an axial keystone graben consisting of two parallel normal faults (Anderson and Christenson 1989). Continuing activity of the uplift is indicated by deformation of late Quaternary alluvial surfaces, fault scarps, and historic earthquakes on the axis of the faulted anticline. Late Pleistocene surfaces are also deformed by the anticline and axial graben. The Sevier River continues to cut through the uplift created over the last 250,000 years. Alluvial sediments of the uplift are being removed. In the uplift, the river is narrow and efficiently shaped. Sediments removed from the uplift are deposited downriver in emphasized meanders.

Near the junction with Utah Highway 12 to Red Canyon and Bryce Canyon, three small basalt flows to the east, a few miles along the Sevier fault, are faulted against Claron Formation (Map 39). The river flows northward through Panguitch Valley about 12 miles to Bear Valley Junction where U.S. 89 joins Utah Highway 20.

Bear Creek

Bear Creek enters the Sevier River from the west (Map 39). Bear Creek is a smaller river system with drainage anomalies just like the Sevier River and East Fork of the Sevier River. Bear Creek is one of the major drainages of the north-central Markagunt plateau. The course of Bear Creek is controlled by structure in places and elsewhere it ignores structure. Bear Creek rises and flows north-northeast in Bear Valley, a north-northeast trending graben floored with alluvium with fault scarps on both sides. The course is consequent in a plunging graben. Bear Creek then flows into lower Bear Valley and leaves the valley through a narrow gorge cut through lava flows and volcanic mudflows. Where Bear Creek cuts this barrier it is neither consequent nor antecedent. Then about halfway along lower Bear Valley the creek makes a sharp turn and flows east out of the Bear Valley graben in a narrow gorge and continues east to join the Sevier River. Before it reaches the Sevier River, it cuts through the middle of a small dome with a radius of 3 miles and 6,600 feet of structural relief. The course between Bear Valley and the Sevier River cannot be consequent and antecedence also seems unlikely. The course must be superposed.

Distinctive gravels of quartzite scattered over high mountains that bound Bear Valley and elsewhere in the northern Markagunt Plateau at least 500 feet above the floor of Bear Valley suggest superposition. The only source of the quartzite is the conglomerate of the Claron Formation, which is not exposed today in any slope of transportation, and so must be a remnant of the alluvial Sevier River Formation that filled valleys and permitted drainages to exit closed valleys. Bear Valley first drains toward the north, then westward in Fremont Wash where the drainage is superposed across a structural high north of Bear Valley.

Bear Creek presently turns abruptly east because of river piracy. Headward erosion of an eastward-flowing tributary of the Sevier River captured Bear Creek. This tributary could also have formed on the Sevier River Formation and become superposed on the structure. Upper Bear Valley is drained by Little Creek that eroded 7.5 miles headward across several structural highs and flows into Parowan Valley. Competition with Bear Creek was won by Little Creek because of higher gradient on Little Creek that allowed capturing more and more of Bear Creek.

The drainage history of Bear Creek has regional implications for the Sevier River. Both Circleville Canyon and Marysvale Canyon on the Sevier River have some similar characteristics. The drainage history of this region is associated with the characteristics of the Sevier River Formation.

The Sevier River Formation is a poorly consolidated, fluvial and lacustrine mudstone, sandstone, and conglomerate that is late Pliocene to early Pleistocene in age. The formation was formerly widely distributed in southwestern Utah, but is now found in isolated patches accumulated in extensional valleys west of the High Plateaus (Kurlish and Anderson 1988). It formed in alluvial fans of poorly sorted pebbles, cobbles, and boulders up to 4 feet in diameter of angular andesite, basalt, and quartzite (Hunt 1956). The lower part of the fan is most poorly sorted, and the upper part contains a few reworked ash beds. The Sevier River Formation is more than 600 feet thick in the Redmond Canyon Quadrangle. One bed in the Aurora Quadrangle to the south yielded a fission track age of 5.2 million years. The formation formerly filled significant parts of the intermontane valleys of the southern high plateaus (Anderson and Rowley 1975; Willis 1991).

Circleville and Marysvale Canyons

A few miles north of Bear Valley Junction, the Sevier enters the narrow Circleville Canyon carved in volcanic rocks of the Tushar volcanic field (Map 40). Circleville Canyon begins in intrusive igneous rocks and extrusive volcanic rocks. The river leaves the igneous rocks and flows into a small patch of Claron Formation. The next bedrock is the Miocene-Oligocene Mount Dutton Formation. The river then reaches a valley with several small normal faults. The Paunsaugunt Plateau has given way to the Sevier Plateau on the east and the Markagunt Plateau and its connection with Tushar volcanic mountains to the west.

Circleville and Marysvale canyons are perplexing geomorphic problems. The three main theories to account for Circleville Canyon are superposition, antecedence, and piracy. The Sevier River, where it flows north through Circleville Canyon, is not a consequent river (Anderson 1987). Here, the Sevier River is in a narrow rock gorge in a structure that forms a transverse barrier across the Sevier Valley. The transverse barrier is beveled by a gently sloping rock surface that represents a long period of stability of the Sevier River. The present river is as much as 500 feet below this surface.

The Sevier River is probably superposed on the rocks and structures in Circleville Canyon (Anderson 1987). Sevier River Formation remnants are exposed on the highlands 1,600 feet above the floor of Sevier River Valley. The bedrock block that Circleville Canyon is cut into has a flat-top rock surface formed during filling of Sevier Valley with Sevier River Formation. Superposition is indicated because the Sevier River Formation is present in the Tushar Mountains and the northern Sevier Plateau high above Marysvale Canyon, and a flat-top erosional rock surface with patches of Sevier River Formation occurs above Circleville Canyon 1,000 feet above the present Sevier River.

The river emerges from Circleville Canyon into a broad, down-faulted valley with a floor of alluvium where the communities of Circleville and Junction are located (Map 40). A few miles north of Circleville, Utah, Highway 62 joins U.S. 89 from Kingston to the east and the East Fork of the Sevier joins the Sevier River from Kingston Canyon.

East Fork of the Sevier River

A major tributary, the East Fork of the Sevier River has an anomalous course (Map 41). It begins in southern Grass Valley and flows north within a down-faulted topographic low until it nears the town of Antimony. There it makes an abrupt, nearly right-angle bend where Otter Creek joins. Otter Creek is a single meandering channel, but many short reaches are an anastomosing

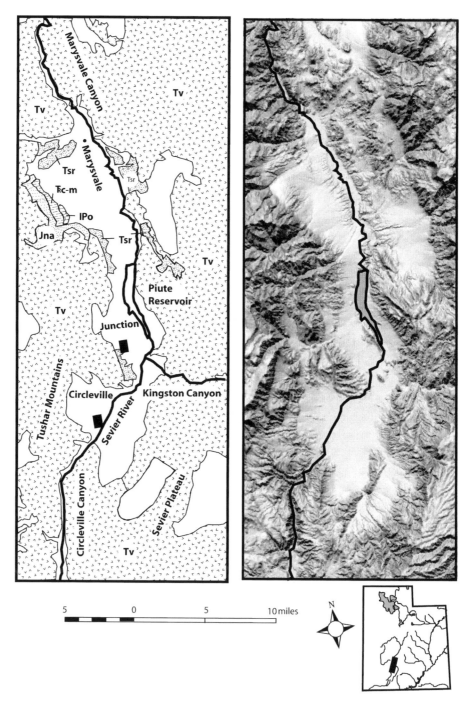

Map 40. Shaded relief and geologic maps of the Circleville and Marysvale Canyons section of the Sevier River. Formation abbreviations are: Tv = Volcanic rocks, Tsr = Sevier River Formation, Jna = Navajo Sandstone, Ṭc-m = Chinle and Moenkopi formations, IPo = Oquirrh Formation.

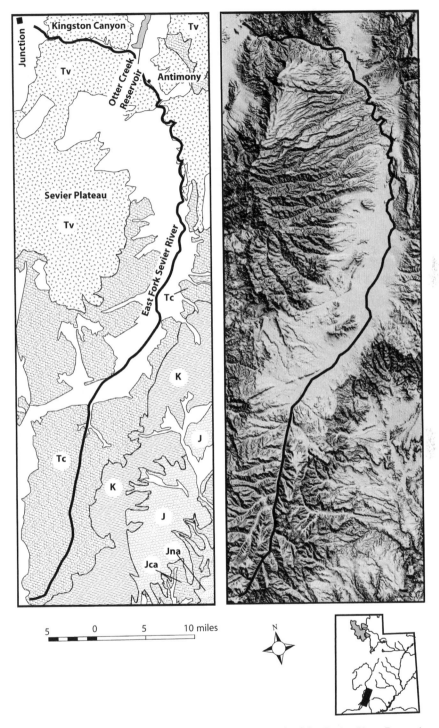

Map 41. Shaded relief and geologic maps of the East Fork of the Sevier River. Formation abbreviations are: Tv = Volcanic rocks, Tc = Claron Formation, K = Cretaceous formations, J = Jurassic formations, Jca = Carmel Formation, Jna = Navajo Sandstone.

channel system with a low width-to-depth ratio, cohesive banks, and low gradients (Dougher and O'Neill 1997).

The East Fork of the Sevier River then flows westward through narrow Kingston Canyon to join the Sevier River near Junction. In Kingston Canyon, the East Fork continues to flow westward and up the dip of bedrock across the tilted fault block of the Sevier Plateau. Rowley (1968) and Rowley, Steven, and Mehnert (1981) describe the development of Kingston Canyon. The explanation for the course of the East Fork must invoke either antecedence or superposition. Kingston Canyon is one of the deepest canyons in the High Plateaus. The East Fork of the Sevier flows westward transversely across the gently east-tilted Sevier Plateau. The Sevier Plateau was uplifted on the Sevier fault by more than 5,000 feet here in the transition from Basin and Range to Colorado Plateau.

The age of igneous rocks in the Kingston Canyon area is crucial to determining the events that have taken place on the river. The Northern Rim rhyolites are 7.6 million years old. A rhyolite dome is 5.4 million years old (Rowley, Steven, and Mehnert 1981). The rhyolite of Forshea Mountain was deposited near Basin and Range faults before uplift of the Sevier Plateau and before cutting of Kingston Canyon. Before the uplift, a river flowed across the site of the Sevier Plateau toward the east-southeast and possibly across the Awapa and Aquarius plateaus to the east (Map 8). The rhyolite of Dry Lake (postdates Forshea Mountain and predates the dome) was deposited during uplift and perhaps before canyon cutting. During uplift the river maintained its course and cut Kingston Canyon (Rowley, Steven, and Mehnert 1981; Anderson 1987).

Superposition is the main control with some river piracy (Anderson 1987). Superposition took place from upper Tertiary valley fill up to 650 feet thick. The rhyolite of Phonolite Hill blocked the river flow, which formed new outlets to the east. Awapa and Aquarius plateaus later were uplifted along faults disrupting the eastern part of the river segment. Topography then took its present appearance with the drainage through Kingston Canyon. There has been little deepening of the river channel since reopening of Kingston Canyon.

Sevier River from Marysvale to Sevier, Utah

The Sevier River continues to flow northward through Piute Reservoir, past Marysvale, and into Marysvale Canyon where it crosses another transverse structure (Map 40). The Tushar Mountains are on the west and the Sevier Plateau on the east. Mount Dutton, 11,035 feet, is the high point on the Sevier Plateau to the east. U.S. 89 continues to follow the Sevier River northward to Junction, Piute Reservoir, and Marysvale.

The Sevier River follows a large alluvial valley to Marysvale. Monroe Peak at 11,226 feet and Marysvale Peak at 10,940 feet are the highest nearby peaks. North of Marysvale, the Sevier enters the narrow Marysvale Canyon carved into Tushar volcanic rocks. The geology is shown on the map of Steven, Morris, and Rowley (1990). Marysvale Canyon is in Bullion Canyon volcanic rocks, Miocene-Oligocene age, in the footwall block of a northeast-striking normal fault. The hanging wall is mostly Mount Belknap Volcanics of Miocene age. A small range of mountains on the east is the Antelope Range with the Sevier Plateau beyond. The Tushar Mountains are to the west.

The Sevier River flows through Marysvale Canyon between the Tushar Mountains with elevations of 12,000 feet and 6,500 feet of relief on the west and the Sevier Plateau with 5,500 feet of relief on the east. Marysvale Canyon gorge, 1,000 feet deep, is cut through a transverse ridge with 1,500 feet of relief similar to Circleville Canyon. The transverse ridge is in the graben between the Sevier fault on the east and the Tushar fault on the west, but faulting is not present on the north and south sides of the ridge. The river flows from Marysvale Valley northward to Sevier Valley. The river meanders through Marysvale Valley, but the river breaks into rapids in Marysvale Canyon because of the elevation difference between Marysvale Valley and Sevier Valley. Alluvium is being removed as the rapids erode headward toward Marysvale Valley. As this process continues, alluvium will eventually be removed from Marysvale Valley.

Marysvale Canyon is not a feature of normal drainage development and could have developed by antecedence, superposition, or river piracy. The river could be antecedent to upraising of the barrier by faulting across the flowing river, or antecedent to regional uplift. The formation of Marysvale Canyon could involve canyon cutting followed by ponding throughout the Sevier Valley, followed by superposition across the barrier and downcutting. The history of the river is coupled with the history of faulting that produced the transverse ridges.

Eardley and Beutner (1934) propose that Marysvale Canyon was formed due to river piracy. The transverse ridge formed an old divide between a north-flowing river and a south-flowing river that drained Marysvale Valley. The lower elevation of Sevier Valley gave the north-flowing river the erosional advantage, and headward erosion across the ridge resulted in capture of the south-flowing river. Some aspects of the drainage pattern around Marysvale such as barbed tributaries suggest piracy and a drainage reversal, but Eardley and Beutner (1934) could find no evidence for a suitable southern outlet to Marysvale Valley. Despite arguments for piracy or antecedence, superposition remains a viable explanation of Circleville, Marysvale, and Kingston Canyons.

Sevier Valley

The river emerges from Marysvale Canyon at the town of Sevier, 10 miles north of Marysvale, and enters the broad, alluvium-floored Sevier Valley (Map 42). Interstate 70 goes west in the valley of Clear Creek that separates the Tushar Mountains to the south from the Pahvant Range to the north. The Sevier River follows Sevier Valley northeastward past Richfield to Salina with the Sevier Plateau on the east and the Pahvant Range on the west. The Sevier Plateau here is Tertiary volcanic rocks that lie on Tertiary and Mesozoic sediments. The Pahvant Range is Tertiary, Mesozoic, and Paleozoic sedimentary rocks thrust on top of Mesozoic sedimentary rocks on the Pahvant thrust fault. This is an important part of the transition from the Great Basin to the Colorado Plateau province. From Richfield to Salina, the Jurassic Arapien Shale is exposed at the valley edge with younger Tertiary volcanic rocks capping the Sevier Plateau.

At Salina, Salina Creek enters from the east. I-70 follows this creek eastward toward Green River, Utah. Salina Creek is the boundary between the Sevier Plateau and the Wasatch Plateau to the north. North of Salina, the Valley Mountains are to the west and the Wasatch Plateau is to the east. Near Gunnison, the Sevier Valley splits, with Sanpete Valley forming the northeast leg and the southern end of Juab Valley (Fayette Valley) forming the northwest leg (Map 43). The San Pitch Mountains separate the two valley segments. The San Pitch River joins the Sevier River near Redmond just 4 miles north of Salina. From Salina to Gunnison, prominent eastern foothills of Arapien Shale are exposed along with Tertiary sediments forming a monocline to the high peaks to the east. Mount Baldy is 10,918 feet, and east of Mayfield is Marys Nipple at 10,984 feet. U.S. 89 follows the San Pitch River northward.

The Sevier River follows Fayette Valley northward with the Valley Mountains on the west and the San Pitch Mountains (also called the Gunnison Plateau) on the east to the Sevier Bridge Reservoir dammed by the Yuba Dam. The Valley Mountains are Tertiary and Cretaceous sediments that are inclined down toward the valley in a monocline. The San Pitch Mountains are Tertiary and Cretaceous sediments that form a steep escarpment on the east.

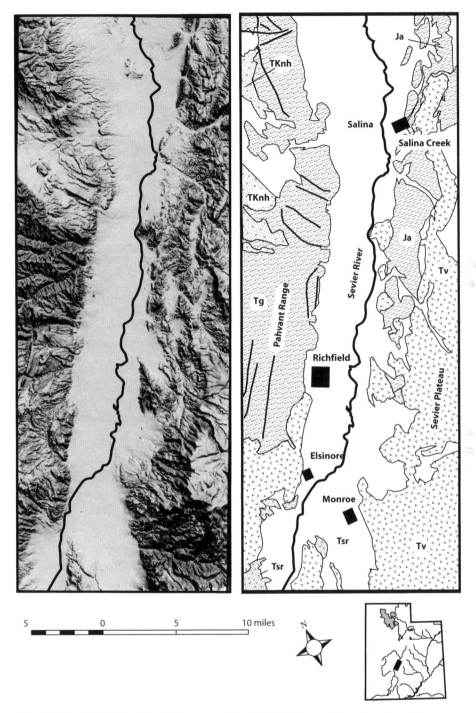

Map 42. Shaded relief and geologic maps of the Sevier Valley section of the Sevier River. Formation abbreviations are: Tv = Volcanic rocks, Tsr = Sevier River Formation, Tg = Green River Formation, TKnh = North Horn Formation, Ja = Arapien Shale.

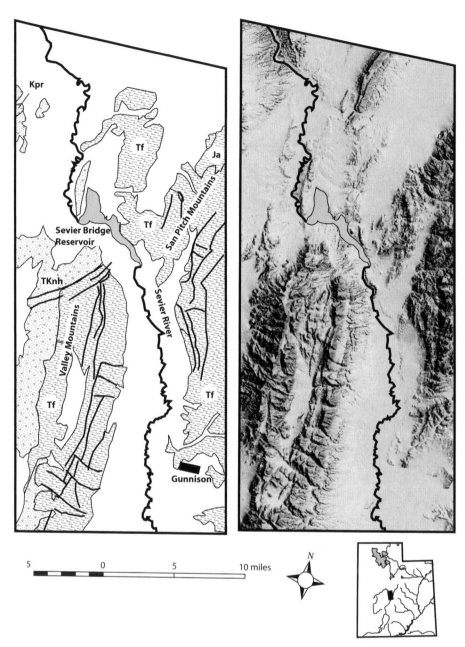

Map 43. Shaded relief and geologic maps of the Sevier Bridge Reservoir section of the Sevier River. Formation abbreviations are: Tf = Flagstaff formation, Kpr = Price River Formation, TKnh = North Horn Formation, Ja = Arapien Shale.

Contrary to evidence offered by system and order in many river systems that indicate that they carve the valleys in which they flow (i.e., the size of a river is proportional to the size of the valley), the Sevier River did not carve some of the valleys in which it flows. The valleys have a more complex origin related to faulting and salt tectonics. The valleys are structural valleys.

The origin of Sevier Valley is complex and controversial. Sanpete and Sevier valleys are the result of complex interaction of the Sevier orogeny thrust faulting and folding, Basin and Range normal faulting, and salt tectonics. Sevier Valley is not a typical valley in the Basin and Range province. The bounding ranges are Tertiary-age rocks, and older Jurassic-age rocks form low hills on the east side of the valley. The controversy involves the extent of Sevier or Laramide compression, folding, and thrust faulting, Basin and Range extension and normal faulting, and rise of salt from the underlying Jurassic-age Arapien Formation. Anderson (1987) suggests that Sevier Valley is a down-faulted valley. Witkind (1992) and Cline and Bartley (2001, 2002a, 2002b) both point out the influence of salt tectonics on the origin of the valley. In fact, it may not be downfaulted at all, but a valley that has formed within an upfaulted mass of salt-bearing Arapien shale.

Cline and Bartley (2001, 2002a, 2002b) propose that Sevier Valley lies on the upthrown but topographically low valley floor footwall of a valley-forming normal fault referred to as the Salina Detachment. The low-angle normal fault puts Tertiary rocks on Jurassic rocks. The downfaulted hanging wall block forms the elevated Wasatch and Sevier plateaus. The Wasatch Monocline is a hanging wall rollover anticline. Arapien Shale has buoyantly rebounded upward in response to unroofing by the Salina Detachment.

Willis (1986) shows a diapir of Arapien on the southeast margin of the valley. The valley fill is Quaternary alluvium and Tertiary beds in a syncline separated from the Arapien diapir by a normal fault that is a west-dipping and possibly diapiric contact.

Structural relations along the southwest Gunnison Plateau suggest that salt diapirism is not operative here (Mattox and Weiss 1987). Drill-hole data along the synclinal axis of the Gunnison Plateau show 1,000 feet of undisturbed salt (see Standlee 1982). The fact that the salt has not flowed suggests that the folding is not a salt-cored anticline (Witkind and Page 1984). In Red Canyon on the northwest margin of the Valley Mountains, there are unconformities between the near-vertical Cretaceous Indianola Group and gently dipping Price River Formation, and between Cretaceous to Tertiary North Horn and Tertiary Flagstaff Limestone. Along the southwest flank of the Gunnison Plateau 4 miles east across Sevier Valley, only one unconformity is present. These relationships confine diapirism to the western margin of northern Sevier Valley and support a nondiapiric origin for

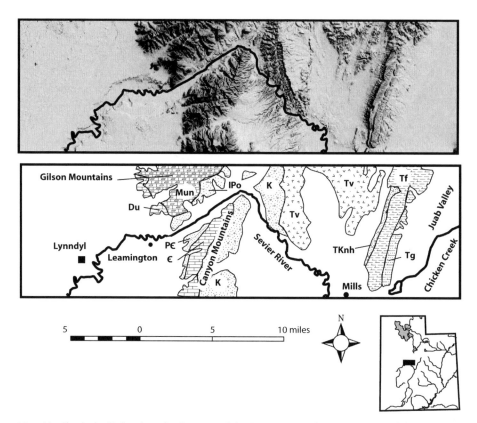

Map 44. Shaded relief and geologic maps of the Leamington Canyon section of the Sevier River. Formation abbreviations are: Tv = Volcanic rocks, Tg = Green River Formation, Tf = Flagstaff Formation, TKnh = North Horn Formation, K = Cretaceous formations, IPo = Oquirrh Formation, Mun = Mississippian formations, Du = Devonian formations, ∈ = Cambrian formations, P∈ = Big Cottonwood Formation.

complex structures including the major monocline along the west front of the Gunnison Plateau. Detailed mapping of Witkind and Weiss (1991) suggests that thrust faulting and high-angle normal faulting rather than diapirs are responsible for the structural complexity.

Mills Valley to Sevier Lake

The Sevier River changes course at the north end of Mills Valley to the northwest and flows across the structural grain to the Great Basin between the San Pitch Mountains and the Valley Mountains (Map 44). Interstate 15 crosses the river just north of the Sevier Bridge Reservoir at the south end of Mills Valley. The Sevier Bridge Reservoir is at the north end of the Valley Mountains. The Canyon Mountains border Mills Valley on the west and the West Hills (west side of Juab Valley) border Mills Valley on the east.

144

The course of the Sevier River was altered at Mills by capture (Oviatt 1987). The capture at Mills is due to lowered base level in the Sevier Desert (tectonic). The age of the capture is much older than Lake Bonneville but much younger than Basin and Range faulting, probably late Tertiary to early Quaternary.

The Sevier River rather abruptly reverses its northward course to flow south into Sevier Lake instead of following an apparently easier course north to Utah Valley. The course of the Sevier River through Leamington Canyon is the result of river piracy and capture at Mills. Mills Gap is near the south edge of the West Hills, a hogback of east-dipping lower Tertiary rocks. Chicken Creek now flows in Mills Gap but is an underfit river, in other words, it is much smaller than the size of the Mills Gap Valley would justify. The precapture Sevier River flowed through Mills Gap into Juab Valley, then north through Juab Valley and across Long Ridge at Currant Creek Canyon into Utah Valley. Currant Creek now flows in Currant Creek Canyon and is also an underfit river. The Sevier River joined the Spanish Fork, Provo, and American Fork rivers and flowed into Ancestral Utah Lake and Ancestral Great Salt Lake.

At the north end of Mills Valley, the Sevier River veers northwest in Sevier Canyon and begins the course of the river around the north end of the Canyon Mountains. Sevier Canyon is cut in Tertiary-Mesozoic sediments to the southwest and Mesozoic sediments and Tertiary volcanic rocks to the northeast. With the capture point at Mills, the river still must make a right-angle bend to flow down Leamington Canyon after flowing through Sevier Canyon. After about 9 miles of Sevier Canyon, the river veers sharply from northwest to southwest where Utah Highway 132 crosses the river at the top of Leamington Canyon. The bend is close to 90 degrees. In Leamington Canyon, the canyon walls are Mesozoic and Paleozoic sedimentary rocks that comprise the Canyon Range. To the north are the Gilson Mountains that are mostly Mesozoic sedimentary rocks. Complicated faulting exposes Pennsylvanian, Permian, and some Precambrian rocks. Hunt (1987) has suggested that the huge canyon, 2,000 feet deep, eroded through deformed Paleozoic formations of the Canyon Range, is the result of superposition rather than capture.

The Sevier River follows Leamington Canyon southwest into the Sevier Desert. The river has left the Colorado Plateau–Great Basin transition zone and entered the Great Basin. The river flows southwest to Delta over a broad and flat plain of the Sevier Desert that is covered with Quaternary sediments that mask underlying geology that includes 5,000 feet of salt possibly brought here from Sevier Valley by an ancestral Sevier River. At Delta the Sevier River enters the Gunnison Bend Reservoir and then heads south to

Deseret. The river veers southwest, skirting a Quaternary basalt flow (Great Stone Face) past Carr Lake and Swan Lake. A final semicircular bend heads nearly due west into Sevier Lake. The Cricket Mountains on the east and the House Range and Black Hills at the south end on the west border Sevier Lake. These ranges are composed of folded and faulted Paleozoic sediments in a Basin and Range setting. Sevier Lake is widely known as a remnant of Lake Bonneville. This is technically true, but the lake had a long history before Lake Bonneville and so did the Sevier River.

The health of the fishery in the main stem of the Sevier River is tied to water released from Piute Reservoir. The river contains rainbow and brown trout from the reservoir (Cook 2000). The Marysvale Canyon mouth marks the boundary between cold- and warm-water fish species. Upstream, the river hosts trout and downriver bass, and nongame fish are present. From Sevier Bridge to Gunnison are 15 miles of river with warm-water fish. From Gunnison to Salina, 14 miles of river host warm-water fish. Good fishing is available in Otter Creek Reservoir. From Sigurd to Sevier Junction, the river hosts brook, brown, and rainbow trout, but irrigation takes much of the water. From Sevier Junction to Panguitch there are 24 miles of tailwater below Piute Reservoir with mainly brown trout. Above Piute Reservoir to Mammoth, 50 miles of river contain brown trout (Demoux 2001). Mammoth Creek, in the Sevier headwaters, is 20 miles of fishing in meadows for 8- to 14-inch brown trout.

The main stem of the Sevier River from Axtell to Yuba Reservoir and from Yuba Reservoir to Leamington is warm water and contains smallmouth bass, walleye, channel catfish, and perch (Prettyman 2001).

The East Fork of the Sevier River contains brook and cutthroat trout upriver from Tropic Reservoir. Below the Tropic Reservoir, insufficient water is available to sustain a fishery. In Black Canyon, brown trout exist with some rainbow and cutthroat trout, and fishing is good between Antimony and Osiris (Prettyman 2001). Kingston Canyon waters contain brown, rainbow, and a few cutthroat trout (Cook 2000; Demoux 2001).

Tributaries of the Sevier River that contain Bonneville Cutthroat trout are Deep Creek, San Stowe Creek, Three Mile Creek, and Manning Creek Reservoir. The fish in remote locations are small (Hepworth and Berg 1997).

The Buenaventura,
Utah's Mythical River

Utah has large rivers, small rivers, and ancient rivers. To complete the story of rivers in Utah, a description of a river that existed only in the minds of mapmakers is appropriate. The myth began with the Domínguez-Escalante expedition that traveled through Utah and adjacent areas in 1776 (Warner and Chavez 1976). The journal kept by Fray Silvestra Vélez de Escalante, together with the map of the journey made by the expedition cartographer Don Bernardo Miera y Pacheco, describes and names the rivers that were crossed. Fray Francisco Atanasio Domínguez was the expedition leader, and his subordinates Escalante and Miera kept the map and journal. Miera was an army engineer, merchant, Indian fighter, government agent, rancher, artist, and cartographer.

The Domínguez-Escalante expedition arrived at the banks of the Green River near Jensen, Utah, on September 13, 1776. The Indians called this river the Seeds-Kee-dee, but the Domínguez-Escalante party named it El Rio de San Buenaventura (St. Goodventure). The Green River had just emerged from Split Mountain Canyon of the South Flank of the Uinta Mountains where it made a large meander bend curving to the north, west, then south. The river continues another 12 miles southwest, then makes another large meander bend known as Horseshoe Bend (Maps 14 and 15, Figure 12). Determination of the regional flow direction of the Green River at this point would have required more exploration than the Domínguez-Escalante party could do. The expedition continued onward to the west and on September 24, 1776, reached Utah Valley and the mouth of the Provo River where it empties into Utah Lake. They went no farther north and so saw neither the Jordan River nor the Great Salt Lake. The journal describes the Spanish Fork River, Hobble Creek (Rio de San Nicolas), the Provo River (Rio de San Antonio de Padua), and the American Fork River (Rio de Santa Ana) that flow into Utah Lake. They learned from the Indians of another lake to the north whose waters were harmful and very salty. The Indians had named the lake Timpanogo.

The expedition turned south and on September 29 camped on the Sevier River near the present town of Mills. The river here, according to the name the Indians had for it, appeared to Escalante to be the Buenaventura. But Escalante doubted that it was the Buenaventura, because it carried much less water than at the crossing near Jensen. Miera, however, just connected the two into a single river flowing into a salt lake (Sevier Lake) that he named after himself, Laguna de Miera, but did not show the western limit of the lake. Had the expedition kept track of elevations, they would have realized that the elevation of the Sevier River near Mills was higher than 4,900 feet and higher than the Buenaventura (Green River) where they crossed near Jensen (4,700 feet).

The map made by Miera showed two rivers entering Laguna de Miera, the Buenaventura and, a bit farther south, the Rio Salado. On Miera's map, the Rio de Yamparicas flowed into the salt lake and one river flowed westward out of it. Miera also, never having seen Great Salt Lake, included it on his map with a large river outlet flowing west off the edge of his map. Evidently Miera was a better artist than mapmaker. The western edge of Miera's map is at the longitude of Sevier Lake. Miera's map is the beginning of the mythical river flowing west from Utah to the Pacific Ocean.

The myth was perpetuated and grew with the next map publication. Baron Alexander von Humboldt was one of the foremost scientists of his time. He drew extensively upon Miera's map and Escalante's journal. He incorporated the Buenaventura, Salado, Yamparicas, and the river flowing west from Great Salt Lake, except he put the Yamparicus flowing into Great Salt Lake from the west, and he did not name Sevier Lake. His map of the western interior of North America, with a geography that was part factual and part mythical, was published in 1808.

The next twenty years saw a confusion of maps showing rivers flowing west out of Utah to the Pacific Ocean. The mapmakers quickly combined Utah Lake and Great Salt Lake into one lake usually called Lake Timpanogos that was drained by a single outlet first called the Multnomah. This river was the heritage of the map made by William Clark from the Lewis and Clark Expedition published in 1814. Lewis and Clark, while ascending the Columbia River, had learned of a river far to the south that flowed north, and named this river the Multnomah. The Multnomah became the river that was the single outlet to Lake Timpanogos. Then John Melish in 1816 drained Sevier Lake with a river between the Buenaventura and San Francisco Bay. So now there was a river that completed the journey from

Utah to the Pacific Ocean. John Robinson, in his map of Mexico, Louisiana, and the Missouri Territory published in Philadelphia in 1819, borrowed from Pike and Humboldt and from William Clark's 1814 map (Morgan 1953). Robinson's map also had one outlet for the prominent Laguna de Timpanogos, but this outlet, named the Timpanogos River, flowed from the lake to San Francisco Bay, rather than Melish's river from Sevier Lake to San Francisco. The following year, 1820, Thompson put the Multnomah back in Lake Timpanogos and the Buenaventura and Salado in Sevier Lake, but continued the Buenaventura to San Francisco. Also in 1820, a French mapmaker, Brué, showed Lake Timpanogos with a river (the Bear?) and a western drain, Rio Yamparicus, but joined Timpanogos with the southern lake called L. Teguayo ou Sale. In 1821, a German mapmaker named Weiland adopted Brué's idea. The Multnomah was gradually removed from Timpanogos. The Vance map of 1826 showed the River Timpanogos arriving at the Pacific between Cape Mendocino and San Francisco, and the Rio de Los Mongos reaching the Pacific south of Cape Orford. The Rio de los Mongos, Timpanogos, and Multnomah all died swift deaths. Map 45, Finley's map of 1826, is an example of maps of that era. Through history the myth of the Buenaventura became a river with a source in Great Salt Lake and then a river almost without a source (Morgan 1947).

The next episode of perpetuation of the myth belongs to the fur trappers. In September of 1824, William H. Ashley was authorized to trade with the Snake Indians west of the Rocky Mountains at the junction of two large rivers, which were supposed to be branches of the Buenaventura and the Colorado of the West. Morgan (1953) suggests this represents the confluence of the Green and Big Sandy rivers. The state of geographic knowledge was represented by the John Robinson map of 1819 (Morgan 1947).

Ashley ascended the Duchesne River, crossed the Uinta Mountains near Bald Mountain to the headwaters of the Bear River, called the Bear River the Buenaventura, and then went on to rendezvous on Henrys Fork (Morgan 1947). Ashley's curiosity was aroused by reports of Grand Lake Buenaventura, because he understood the Indians to say a river flowed from its western side into unknown country. The *Missouri Advocate* newspaper of March 11, 1826, printed momentous news of a "New Route to the Pacific Ocean, discovered by General William H. Ashley, during his late expedition to the Rocky Mountains." The route, after leaving St. Louis, followed the north side of the Missouri River, then pursued the Platte to its source, crossed the headwaters of the Rio Colorado of the West, struck a north-south chain of mountains with a wide gap in it, and finally followed the Buenaventura to the Pacific Ocean. Ashley reported that the Buenaventura

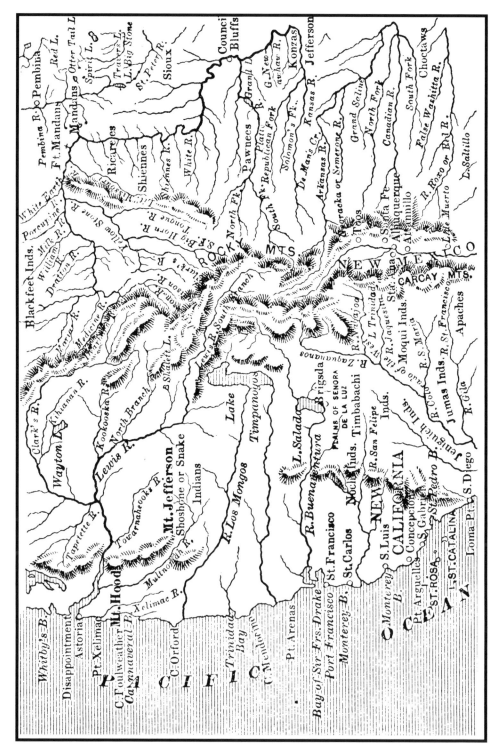

Map 45. Finley's map of 1826 showing the location of Lake Timpanogos, Lake Salado and the Rio Buenaventura (from Bancroft 1889).

flowed into a large lake and at the lake's west end a river flowed out. What lay west and south of Great Salt Lake was a great mystery. Still, the idea of a westward-flowing river did not die.

Jedediah Smith, Joe Walker, and John C. Fremont all searched for the Buenaventura River. Smith became a proficient trapper, an adept trader, and an accomplished explorer in the short span of his life. He was completely inexperienced when he was recruited by William Ashley and Andrew Henry in 1822, and he was killed by Indians on the Santa Fe Trail in 1831 at age thirty-three. Smith began his search for the Buenaventura as a possible western outlet to Great Salt Lake in 1826. Smith sent a party of four men in bull boats under the leadership of David E. Jackson to explore the north and west shorelines of Great Salt Lake by boat. After twenty-four days of cruising the coastline, they found no outlet (Goetzmann 1966), though they reported seeing where it might have been. The south and west shorelines remained unexplored until Smith's great southwest expedition. With fifteen men, Smith explored the region south and west of Great Salt Lake including the Sevier and Virgin Rivers, crossed the Mojave Desert, and reached the Pacific coast at San Diego on December 12, 1826, with no sign of the Buenaventura observed (Morgan 1953). Smith saw the impressive Sierra Nevada Mountains through which the Buenaventura must pass to reach the Pacific. He considered the Sacramento River a possible candidate for the Buenaventura. Another choice was a tributary of the San Joaquin. Jedediah Smith applied the name Buenaventura to the Sacramento River, but it is doubtful that, after seeing the Sierra Nevada Mountains, Smith envisioned it as a river arising far to the east. Trapping beaver up the Sacramento to the head of Sacramento Valley and beyond into the Sierra confirmed that these two rivers arose in the Sierra Nevada and not far to the east.

The next chapter belongs to the French-born American soldier, explorer, and fur trader Captain Benjamin Louis Eulalie de Bonneville. With a leave from the army, private backing, and 110 men, he set out to explore the West in 1832 (Irving 1850). Curiously, Bonneville never reached the Great Salt Lake or the Bonneville Basin. In 1833, Bonneville sent Joseph Redford Walker to the Great Salt Lake to explore the northwest and west sides of the lake. This expedition produced Bonneville's map (Map 46) where Great Salt Lake was named Lake Bonneville. The map showed no outlet to the west. The map shows a river named Buena Vista arising in the Sierra Nevada (California Mountains) and flowing into San Francisco Bay (Sacramento River) (Miller 1980).

The myth was finally laid to rest when the exploration of John C. Fremont completely circumnavigated the entire Great Basin (he named the Great Basin) and found no river crossing the basin. Fremont arrived in Utah

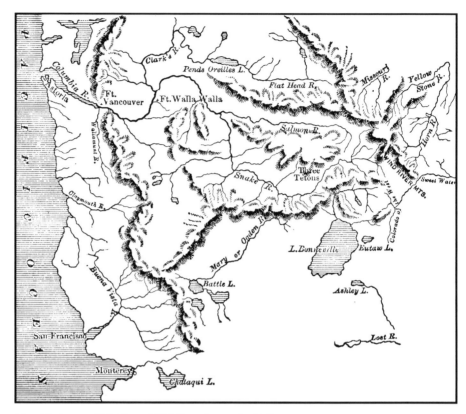

Map 46. Bonneville's Map of 1837 (from Bancroft 1889).

Valley on May 25, 1844, after he had completed a circuit of the Great Basin 12 degrees in north-south diameter and 10 degrees in east-west diameter. The circuit was completed in eight months since leaving Great Salt Lake, and in 3,500 miles of travel no evidence of the Buenaventura River was found.

Smaller Utah Rivers
of the Colorado Plateau

The Colorado River and the Green River tend to dominate thinking about rivers on the Colorado Plateau. However, smaller tributaries with their own special significance fed these two great rivers.

The early Tertiary lakes lasted until uplift of the Colorado Plateau region during the Oligocene. Overflow of the lakes was southward forming the beginning of the Green River and Colorado River systems in Utah. Drainage that had been centripetal into the lakes became radial westward and southward away from the lakes. The westward drainage was superimposed across Laramide structures of the faulted boundary between the Colorado Plateau and the Great Basin. Some of these superimposed courses became antecedent to the rising Wasatch Range.

The Colorado Plateau west of the Colorado River was raised to form the High Plateaus consisting of up-faulted plateaus and downfaulted basins. The High Plateaus form the drainage divide between the Great Basin and the Colorado Plateau. Long tributaries flowing from the high plateaus to the Colorado River include the Paria, Escalante, Fremont, Muddy, San Rafael, and Price rivers. The Fremont and Muddy join to form the Dirty Devil.

Salina Creek, which heads at a wind gap against a tributary of the Muddy River, flows to the Great Basin after cutting deeply into the plateau. Salina Creek is anomalous; it appears to follow the northern edge of volcanic rocks forming faulted masses of mountains that include Hilgard, Terrel, and Marvine. Salina Creek could have begun as a strike valley curving west and around the north base of the volcanic rocks. Salina Creek was antecedent across north-trending fault blocks.

The high plateaus end northward at the Spanish Fork River. The Spanish Fork River flows west and the White River fork of the Price River flows

Map 47. Shaded relief and
geologic maps of the Paria
River. Formation abbrevia-
tions are:
Tc = Claron Formation,
K = Cretaceous formations,
J = Jurassic formations,
Jca = Carmel Formation,
Jna = Navajo Sandstone,
Ŧc = Chinle Formation,
Ŧm = Moenkopi Formation,
Pk = Kaibab Formation.

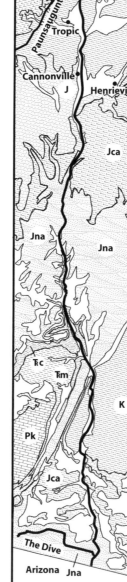

5   0   5 miles

N

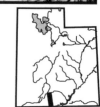

east. Both of these rivers occupy strike valleys around the plunging end of the anticline that forms the north end of the Wasatch Plateau (Map 8).

The drainage east and west of the High Plateaus differs. Several long parallel tributaries of the Colorado began forming during the Oligocene when Green River Lake overflowed, and the Green and Colorado rivers began their courses south across Utah. Drainage west from the high plateaus also began during the Oligocene. South of Mount Nebo, drainage was guided into north-south graben valleys forming a rectangular pattern in the developing Basin and Range province.

Little or no significant bedrock incision is apparent in Canyonlands and the High Plateaus sections of the Colorado Plateau during the last 10,000 years. Most of the rivers are aggrading, but historical records show many rivers have been downcutting since the end of the nineteenth century.

At several places on the Colorado Plateau, canyon cutting was interrupted and rivers became superimposed on geologic structures outside their former valleys. Halls Creek occupies a 35-mile-long strike valley on the east side of the Waterpocket Fold. At two places superimposed Halls Creek leaves the strike valley and meanders into 500-foot-deep canyons in resistant sandstone.

Paria River

The Paria River was named in John Wesley Powell's era by the Mormon scout and explorer Jacob Hamblin, who named it the Paria from a Piute Indian word meaning "muddy water" (Van Cott 1990), which it certainly is especially following a summer cloudburst. The river is eroded on the upthrown side of the Paunsaugunt fault in Cretaceous formations (Map 47). The western tributaries of the Paria flow from exposures of the Tertiary lake sediments on the high Paunsaugunt Plateau and erode the cliffs that border the eastern Paunsaugunt Plateau, producing the spectacular scenery of Bryce Canyon National Park (Figure 19).

Headward erosion into the plateau may eventually capture the headwater of the East Fork of the Sevier River. The Paria then heads southeastward through Tropic Valley toward its confluence with the Colorado River. The river soon cuts through Cretaceous rocks into Jurassic and Triassic rocks on the north end of the Kaibab Uplift in cliff-walled canyons in the Navajo Sandstone. The Paria River then cuts through the prominent East Kaibab monocline and flows back into the Cretaceous rock sequence. The river continues down the stratigraphic section into Jurassic rocks and leaves Utah in a steep canyon carved through Navajo Sandstone.

Figure 19. Bryce Canyon, Utah. Bryce Canyon formed in the lake deposits of the Claron Formation at the headwaters of the Paria River. Author photograph.

Beaver River

The Beaver River begins on the southwest flank of the Tertiary-age Marysvale volcanic field. The lower part of the deep Beaver River gorge provides a spectacular cross section of details of the build-up of this volcanic field. Basin and Range faulting that began in the Miocene and progressed to the Recent elevated the Tushars to their present relief, and led to downcutting of the Beaver River.

Birch Creek, North Fork Creek, Pine Creek, and Briggs Creek, tributaries of the Beaver River, contain Bonneville cutthroat trout (Hepworth and Berg 1997).

Virgin River

The North Fork and East Fork of the Virgin River, each with drainage areas of 348 square miles, originate in the Pink Cliffs of the southern Markagunt Plateau. Average elevation of the two basins is 7,350 feet. The two forks join downriver from Springdale, Utah, cross the Hurricane fault, and enter the Basin and Range province.

Long Valley Junction is the drainage divide between the Sevier River flowing north and the East Fork of the Virgin River flowing south. Both of these rivers flow along the general trace of the Sevier fault. The East Fork of the Virgin River begins in Tertiary sedimentary rocks and heads southwest

Figure 20. The Virgin River in Zion Canyon viewed from the East Rim overlook looking south. The Virgin River is deeply incised into the Navajo Sandstone. Author photograph.

through Tertiary, Cretaceous, and Jurassic strata of the Grand Staircase (Map 48). At Mount Carmel Junction, the East Fork of the Virgin abruptly turns west as it flows into the narrow Parunoweap Canyon in Navajo Sandstone. Before reaching Springdale, the East Fork has incised down to the Kayenta and Moenave formations. Just south of Springdale, the North Fork of the Virgin, which flows through Zion Canyon, joins the East Fork in the Chinle Formation.

The alluvium along the East Fork preserves a 1,600-year record with evidence of large or frequent floods. Flood episodes occurred at about 1,600 years ago, 1,000 years ago, and in the past 300 years. The largest floods on the East Fork occurred in the historic period (Enzel, Webb, and Benito 1997).

The North Fork of the Virgin River begins south of Navajo Lake on the southern Markagunt Plateau and flows from the Pink Cliffs of Tertiary formations through the Cretaceous Grey Cliffs, heading southwest and then west into Jurassic Carmel and Navajo Sandstone of the White Cliffs (Map 49). Upon entering the Navajo Sandstone, the river turns southeast following the northwest-southeast joint set that controls the development of the Zion Narrows and so many of the rivers and canyons in Zion National Park. The river leaves the narrows, flows southward in Zion Canyon, and cuts down into Triassic beds. The North Fork of the Virgin River carved the cliff-walled Zion Narrows and Zion Canyon in Navajo Sandstone (Figure 20). The Navajo Sandstone is strong enough here to stand in 1,000-foot-high cliffs in the slot canyon that is barely 16 feet wide at the bottom.

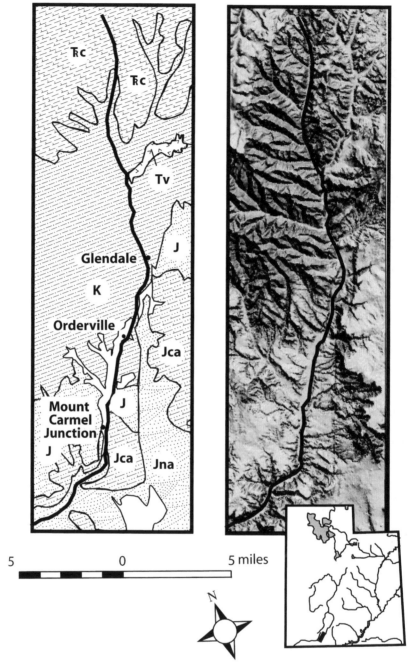

Map 48. Shaded relief and geologic maps of the East Fork of the Virgin River. Formation abbreviations are: Tv = Volcanic rocks, K = Cretaceous formations, J = Jurassic formations, Jca = Carmel Formation, Jna = Navajo Sandstone, ℟c = Chinle Formation.

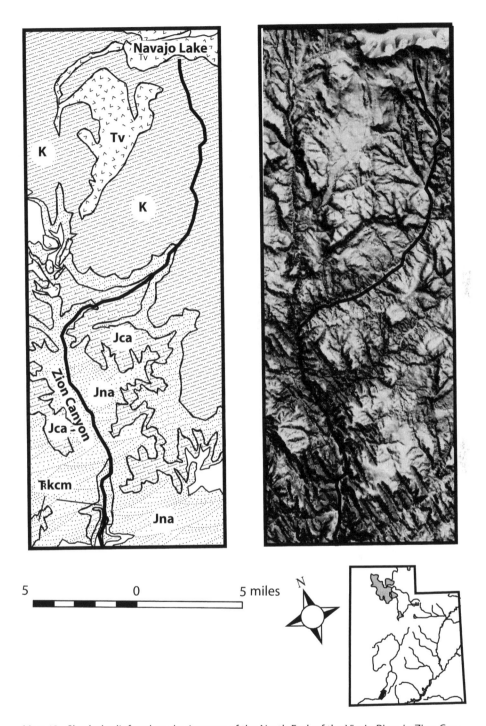

Map 49. Shaded relief and geologic maps of the North Fork of the Virgin River in Zion Canyon. Formation abbreviations are: Tv = Volcanic rocks, K = Cretaceous formations, Jca = Carmel Formation, Jna = Navajo Sandstone, Ŧkcm = Kayenta, Moenave, and Chinle formations.

Figure 21. Waterfall following a thunderstorm, flowing from a hanging canyon within Zion Canyon. Incision of the Virgin River in the Navajo Sandstone has left this tributary canyon as a hanging valley. Author photograph.

Slot canyons are narrow and deep channels in bedrock. They form when river downcutting greatly exceeds river valley widening. The Colorado Plateau has many slot canyons because of its relatively recent uplift and accompanying river incision. The slot canyon at the Zion Narrows, one of the longest slot canyons, follows the prominent northwest-trending joint set that determines the direction of many of the smaller rivers in Zion National Park. The steep gradient of 50 to 80 feet per mile and averaging 71 feet per mile in Zion Canyon caused the Virgin to deepen its canyon more rapidly than mass wasting widened the canyon. The Zion Narrows slot canyon likely formed in the last one million years by erosion of a few inches of bed-

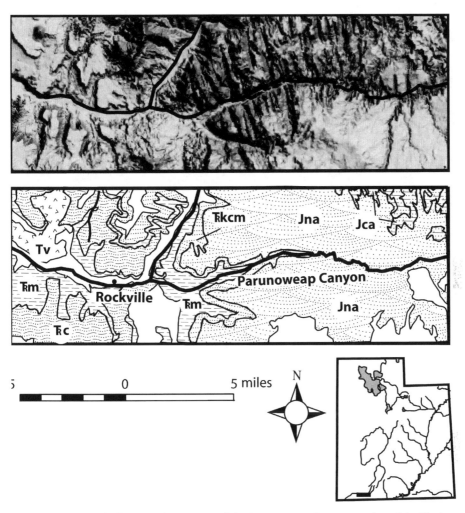

Map 50. Shaded relief and geologic maps of the Parunoweap Canyon section of the Virgin River. Formation abbreviations are: Tv = Volcanic rocks, Jca = Carmel Formation, Jna = Navajo Sandstone, ₹c = Chinle Formation, ₹kcm = Kayenta, Moenave, and Chinle formations, ₹m = Moenkopi Formation.

rock during each of the largest floods (Andrews 1997). The river removes more than 1 million tons of rock from Zion each year. Tributary rivers with much smaller discharge could not keep pace with the canyon deepening, and their canyons are left as hanging valleys that only contain water during rainstorms or rapid melting of the snow, which produces spectacular waterfalls (Figure 21).

After the confluence of the East and North forks, the Virgin flows west past Rockville and Virgin in the Moenkopi Formation with prominent exposures of basalt flows on the north bank (Map 50). The river is incised into the Permian Kaibab Formation on the footwall of the Hurricane fault (Map

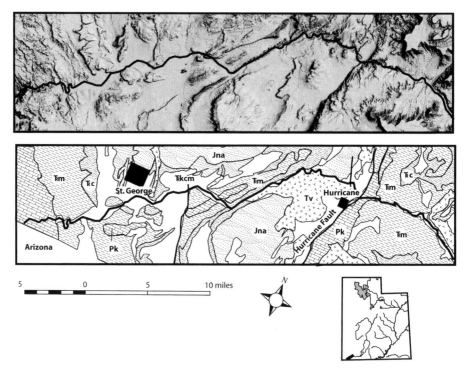

Map 51. Shaded relief and geologic maps of the Hurricane-St George section of the Virgin River. Formation abbreviations are: Tv = Volcanic rocks, Jna = Navajo Sandstone, Ŧc = Chinle Formation, Ŧkcm = Kayenta, Moenave, and Chinle formations, Ŧm = Moenkopi Formation, Pk = Kaibab Formation.

51). The river crosses the fault and turns southwest in Moenkopi Formation and Chinle Formation toward St. George and the Arizona border. The river leaves Utah in the Kaibab Formation.

Nowhere in the west are there better examples of the processes that widen river valleys than in Zion Canyon. The high, nearly vertical cliffs of Navajo Sandstone and underlying softer Moenave, Chinle, and Moenkopi formations lead to rock falls and landslides, both considerable hazards to residents and visitors to the area. On October 18, 2001, a 200- to 300-ton slab of conglomerate of the Shinarump Member of the Chinle Formation destroyed a home in Rockville. Between Rockville and Zion Park, numerous similar blocks attest to the hazard there. Springdale, on the Virgin River at the entrance to Zion Canyon, has large landslides on the northeast and southwest sides, and a large landslide blocked the road in Zion Canyon and dammed the Virgin River on April 12, 1995 (Rowley et al. 2002).

The Virgin River is a perennial river. The present elevation and course of the river result from extensive bedrock cutting in the last 1 to 2 million years controlled by Quaternary movement on the Hurricane fault of 2,000 to 2,800 feet. The river was within 65 to 100 feet of its present elevation by 293,000 years ago.

Underlying bedrock controls the width of the river at Rockville. The river is wide in the Moenkopi, Chinle, Moenave, and Kayenta formations, and narrow in Navajo Sandstone. The Virgin River was dammed by the Crater Hill basalt flow near the ghost town of Grafton about 100,000 years ago. The dam formed Lake Grafton, which covered about 10 square miles. Erosion has removed most of the 300 feet of sediment that accumulated in Lake Grafton. Sediment in the lake was deposited in climate conditions similar to today (Sorensen and Lohrengel 2001).

The Virgin River in Zion National Park has changed from a wide, braided river to a narrower, meandering single channel in the 1900s. The shift in the character of the river could be the result of a change in the pattern of flooding (Sharrow 2004). Flooding and the increased sediment brought to the river could have produced the braided river pattern.

Hereford, Jacoby, and McCord (1996) identify, date, and interpret patterns of erosion and deposition along the Virgin River over the past 1,000 years. Widespread terraces and the active channel floodplain are prehistoric, settlement, historic, or modern. Entrenchment of the river took place in prehistoric and historic time.

Native fish populations in the lower Virgin River have declined due to alteration and loss of habitat from mining, agriculture, and urbanization, and due to introduction of the nonnative red shiner (Fridell and Morvilius 2005). The river contains native Virgin spine dace, speckled dace, Virgin River chub, desert sucker, woundfin, and flannelmouth sucker. Low numbers of the nonnative largemouth bass, black bullhead, and green sunfish are found. Virgin River tributaries that contain Bonneville cutthroat trout include Reservoir Canyon Creek, Water Canyon Creek, Leap Creek, South Ash Creek, and Leeds Creek. Some of these fish are remnant native populations and some have been transplanted (Hepworth and Berg 1997). Trout are present in low numbers in the Virgin River and are found occasionally in Zion Canyon.

Escalante River

The Escalante River flows southeast from Utah into Lake Powell as it skirts the western margin of the Circle Cliffs Uplift. To the north, it bends westward where it crosses and then follows the Escalante monocline.

The Escalante River was the last river to be discovered and added to maps in the United States. It is also considered one of the most crooked rivers in the United States. Gregory measured a 14-mile, straight-line segment

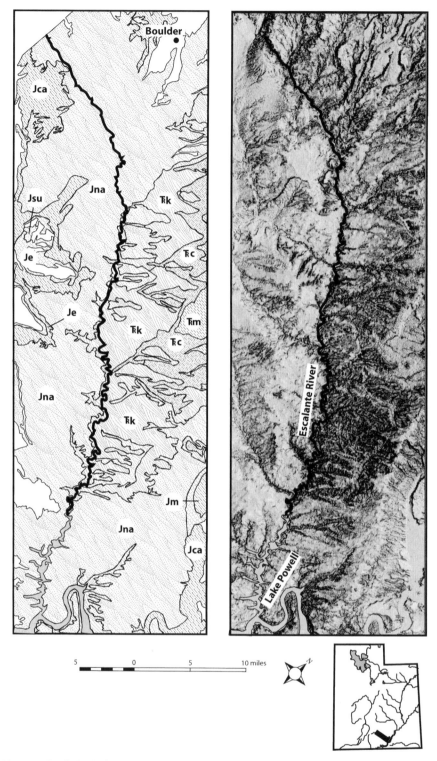

Map 52. Shaded relief and geologic maps of the Escalante River. Formation abbreviations are: Jm = Morrison Formation, Jsu = Summerville Formation, Jca = Carmel Formation, Je = Entrada Formation, Jna = Navajo Sandstone, Ŧk = Kayenta Formation, Ŧc = Chinle Formation, Ŧm = Moenkopi Formation.

that included 35 miles of river. Almon Thompson, Powell's brother-in-law, named the river (Van Cott 1990).

North Creek, Birch Creek, and Upper Valley Creek flow east from the Table Cliffs Plateau and the Escalante Mountains, joining to form the east-flowing Escalante River a few miles west of Escalante, Utah (Map 52). The river soon enters a canyon east of Escalante in Navajo Sandstone, but even though it is confined to steep canyon walls, the river continues to meander as it gradually turns southeast between the Straight Cliffs that border the Kaiparowits Plateau on the southwest and the Circle Cliffs on the northeast. Hydrocarbon fluids permeated the Navajo Sandstone here, removed much of the red iron oxide coloring matter, and bleached the sandstone. The river remains in the Navajo Sandstone canyons.

At the confluence with Horse Canyon, the river cuts into the Kayenta Formation at river level, but the canyon is still a steep-walled, narrow canyon between Navajo Sandstone cliffs. The river has continued to downcut into the Wingate Sandstone. Meanders grow larger as the river flows southeast through Escalante Canyon to Lake Powell. Just before entering Lake Powell, the river has cut into Chinle Formation with cliffs of Wingate, Kayenta, and Navajo formations.

The modern Escalante River has been shaped by aggradation with sediment from Straight Cliffs landslides, by overlying alluvial-fill flood deposits, and by historic floodplain deposits (Patton et al. 1991). The Escalante River has been aggrading for most of the last 10,000 years, but there were two late periods of rapid incision at 2,500 to 1,900 years BP and 1,000 to 300 years BP (Patton and Boison 1986). Two terraces are visible in the lower reaches of Harris Wash, representing incision of that Escalante tributary.

The Escalante River is extremely remote with minimal access for fishing in deep, narrow sandstone canyons. Flash floods are frequent and summer water temperatures are high. Fridell, Morvilius, and Wheeler (2004) and Morvilius and Fridell (2005) describe the distribution of fish. The Escalante River contains six species of native fish: speckled dace, flannelmouth sucker, bluehead sucker, roundtail chub, mottled sculpin, and Colorado River cutthroat. Of these the speckled dace is most abundant. Colorado River cutthroat are found only in the headwaters of some tributaries. Seven nonnative species are present: cutthroat trout, brown trout, brook trout, fathead minnow, channel catfish, common carp, and red shiner. Cutthroat trout were stocked in Calf Creek in 1977 and remain at the original stocking point in the Escalante River only at the confluence with Calf Creek. Brown trout were stocked in Calf Creek and Pine Creek in 1967. Abundant brown trout are present now in lower Calf Creek and Deer Creek. Cutthroat trout were also found in Calf Creek. Brown trout were also found

Map 53. Shaded relief and
geologic maps of Muddy Creek.
Formation abbreviations are:
Tf = Flagstaff Formation,
TKnh = North Horn Formation,
Kpr = Price River Formation,
Km = Mancos Shale,
Kdcm = Dakota Sandstone and
Cedar Mountain Formation,
Jm = Morrison Formation,
Jscu = Summerville and
Curtis formations,
Je = Entrada Formation,
Jca = Carmel Formation,
Jna = Navajo Sandstone,
Ʀkwc = Kayenta, Wingate,
and Chinle formations,
Ʀm = Moenkipi Formation,
Pk = Kaibab Formation.

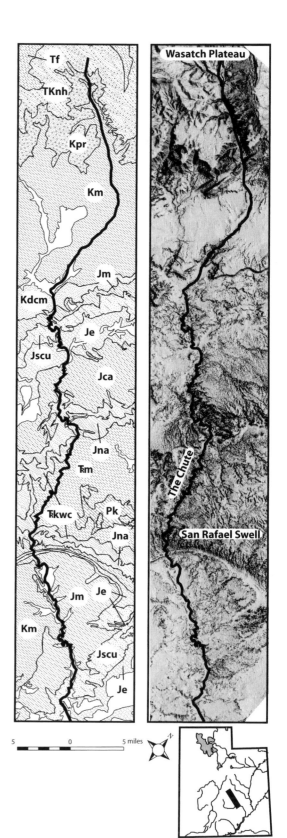

in Sand Creek, Boulder Creek, Calf Creek, Deer Creek, and Mamie Creek. Rainbow trout were stocked in 1941 in the east forks of Deer Creek and Boulder Creek, but none remained in 2003. Native fish populations are present from above Calf Creek down to Boulder Creek. The lower Escalante contains nonnative channel catfish, carp, flathead minnow, red shiner, yellow bullhead, striped bass, largemouth bass, and green sunfish that migrate upriver from Lake Powell

Muddy Creek

Muddy Creek begins on the southern high Wasatch Plateau and flows southeast and straight into a steep-walled canyon through the Cretaceous sedimentary section that once covered the San Rafael Swell and much of the Colorado Plateau (Map 53). These sediments are exposed on the Book Cliffs to the north, and the exposure continues southward as the eastern escarpment of the Wasatch Plateau. Muddy Creek flows from the Wasatch Plateau into the southern end of Castle Valley and turns south to cross Interstate 70 and begins to meander southeast across the San Rafael Swell. The river cuts into the core of the San Rafael Swell into the oldest rocks exposed there.

The San Rafael Swell is an anticlinal uplift produced during the Laramide mountain-building event. The present expression of the swell is a topographic high that is 2,000 feet higher than the adjacent San Rafael Desert or Castle Valley. The Price, San Rafael, and Muddy rivers cross the San Rafael Swell. All are probably superimposed (Davis 1901:140; Gilluly 1929).

Muddy Creek has cut the deepest canyon on the San Rafael Swell, known as the Chute (Figure 22). This deep slot canyon is carved through the Triassic Moenkopi Formation into Permian Black Dragon Dolomite (Kaibab of older reports) and White Rim Sandstone (Coconino Sandstone of older reports). The river flows in a steep-walled, narrow slot canyon through the San Rafael Reef through Navajo Sandstone, Carmel, and Entrada formations, and then meanders southeast through Summerville, Curtis, and Morrison formations to the confluence with the Fremont River.

Fremont River

The Fremont River begins as Lake Creek flowing north out of Widgeon Bay of Fish Lake beneath Fish Lake High Top Plateau, part of the Awapa and Sevier plateaus (Map 54). The plateau is capped by volcanic rock. Fish Lake occupies a graben formed by northeast-to-southwest-striking normal faults with about 1,000 feet of displacement.

Webber and Bailey (2003) suggest that Fish Lake was originally without

Figure 22. The Chute of Muddy Creek. The deepest canyon in the San Rafael is carved through the Triassic Moenkopi Formation (Sinbad Limestone and Black Dragon Members) into the resistant Permian limestone and sandstone cliffs (Black Dragon Dolomite–Kaibab of older reports, and White Rim Sandstone–Coconino Sandstone of older reports) of the slot canyon. The steep cliffs form a narrow map pattern of outcrop that cannot be shown on the geologic map. Photograph courtesy of Gary Colgan.

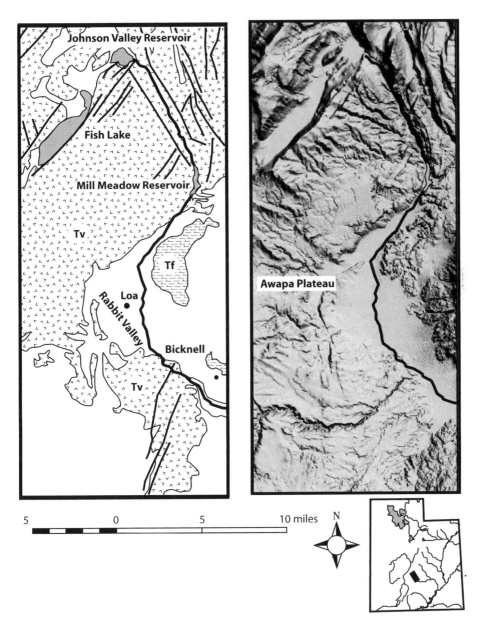

Map 54. Shaded relief and geologic maps of the upper Fremont River from Fish Lake to Bicknell. Formation abbreviations are: Tv = Volcanic rocks, Tf = Flagstaff Formation.

an outlet and was captured by the Fremont River. River terraces 394 feet above the river may have formed before the capture. Lake Creek flows into Johnson Valley Reservoir and south out of the reservoir with no meanders between two faults that bound a northwest-to-southeast-trending graben. The graben is filled with sandstone and conglomerate that have been

169

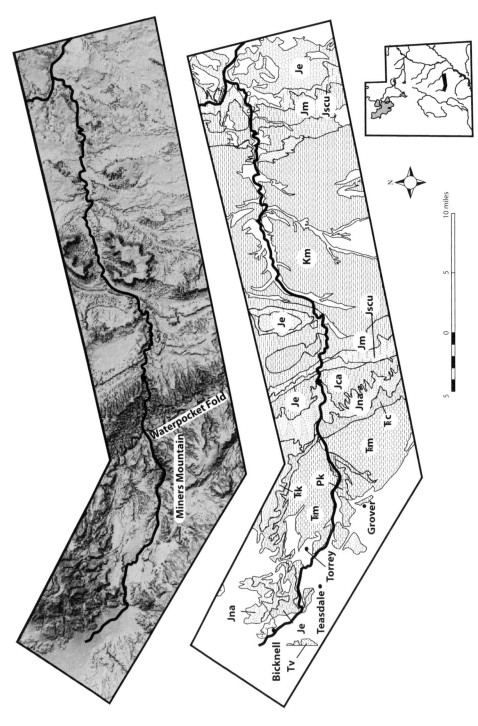

Map 55. Shaded relief and geologic maps of the Fremont River from Bicknell to the confluence with Muddy Creek. Formation abbreviations are: Tv = Volcanic rocks, Km = Mancos Shale, Jm = Morrison Formation, Jscu = Summerville and Curtis formations, Je = Entrada Formation, Jca = Carmel Formation, Jna = Navajo Sandstone, Ͳk = Kayenta Formation, Ͳc = Chinle Formation, Ͳm = Moenkopi Formation, Pk = Kaibab Formation.

eroded and then capped by a volcanic boulder surface that is now about 492 feet above the Fremont River. The cosmogenic helium-3 (^{3}He) exposure age of pyroxene from the volcanic rock boulders ranges from 700,000 to 900,000 years and provides a constraint on the length of time needed for the Fremont River to cut down to the present level (Dickson et al. 2006). Just downriver from Johnson Valley Reservoir, a series of very high river terraces are preserved that are capped by volcanic boulders. The exposure age of boulders on one terrace 508 feet above river level determined from the concentration of helium-3 (^{3}He) range from 607,000 to 508,000 years (Marchetti, Cerling, and Waitt 2005).

The river flows to Mill Meadow Reservoir, then southwest in volcanic rock into Rabbit Valley, Loa, Bicknell, and Bicknell Bottoms. Bicknell Bottoms is a marshy area where the Fremont River has deposited much sediment on the down-thrown side before crossing to the upthrown side of the Thousand Lake fault zone. The Teasdale fault also intersects the river here. The Thousand Lake fault represents the easternmost limit of the transition from the Basin and Range to the Colorado Plateau physiographic province.

The river then heads east to Torrey and Capitol Reef between high cliffs of Wingate Sandstone and, in the distance, Navajo Sandstone (Map 55 and Billingsley, Huntoon, and Breed 1987). Thousand Lake Mountain is on the north and Boulder Mountain is on the south. Both are capped with volcanic rock. Between Torrey on the west side of the Waterpocket monocline and the Dirty Devil River near Hanksville, the Fremont River descends nearly 2,000 feet. The slopes nearer the river are Chinle and Moenkopi formations. The river flows east then southeast on alluvium covering the Moenkopi Formation. The Fremont then heads east into the Miners Mountain anticline and cuts down into the Kaibab and Cutler formations in the core of the anticline (Figure 23). River terraces are preserved here.

The Fremont River is joined by tributary Sulphur Creek from the west. Sulphur Creek is characterized by goosenecks, pronounced entrenched meanders representing superimposition of a meandering river on the underlying rock structure. Sulphur Creek Canyon at the Goosenecks exposes Permian-age Cutler Group Sandstones 250 to 290 million years old.

The Fremont emerges from Miners Mountain at Fruita and flows back into the Moenkopi and Chinle formations. The Fremont then enters a cliff-walled canyon as it cuts through the Waterpocket Fold at nearly right angles with significant meanders amid cliffs of Wingate, Kayenta, and Navajo Sandstone. Along the Fremont River, high-elevation, gently sloping bedrock surfaces at the base of steeper slopes are covered with gravel that may be glacial outwash, landslide debris, or fluvial sediment. Modern rivers are deeply incised into old bedrock surfaces and have formed numerous terraces. One

Figure 23. The Fremont River as it cuts east through the Miners Mountain uplift. View to the west with Boulder Mountain in the distance. Author photograph.

prominent terrace along the Fremont River that extends from the flanks of the Aquarius Plateau through Capitol Reef increases from 50 feet to 230 feet above the river elevation at Fruita possibly due to displacement on the Thousand Lake fault.

Terrace systems can be correlated along the Fremont and Dirty Devil rivers down to the confluence with the Colorado River. Downcutting since the lowest river terrace was formed is 164 feet at the Colorado River, 100 feet at Hanksville, 200 feet at Cainville, and 550 feet on the flanks of Boulder Mountain and Thousand Lake Mountain (Howard 1986). Terraces covered with large, dark-colored boulders are prominent along the Fremont River. The boulders are volcanic rock debris flows derived from the volcanic caps on Boulder Mountain and Thousand Lake Mountain that rise high above the river and Capitol Reef National Park. The Fremont River has cut down into bedrock from 66 to 850 feet below these surfaces (Waitt 1997). The length of time that these boulders have been exposed to cosmic rays, representing their exposure age and hence the time elapsed since their emplacement and the time required by the river to reach its present level, has been recently determined (Marchetti 2002; Marchetti, Cerling, and Waitt 2000; Marchetti and Cerling 2005). The river terraces capped with debris flows have ages that show no correlation or age relationships that can be correlated with glaciation on nearby high mountains. The Johnson Mesa terrace is 300 feet above the present Fremont River and has a helium-3 (^{3}He) exposure age of 180,000 to 205,000 years. Incision rates on the Fremont River calculated from exposure ages are about 17 inches per 1,000 years. Beryllium-10 (^{10}Be) ages of the three highest terraces on the Fremont River are 60,000, 102,000, and 151,000 years (Repka, Anderson, and Finkel 1997).

The river emerges from the Waterpocket Fold amid east-dipping Carmel, Entrada, Curtis, Summerville, and Morrison formations and begins a meandering course on Mancos Shale northwest toward Cainville and Hanksville. At the Cainville Reef, the Ferron Sandstone Member of the Mancos Shale deflects the meandering river northward. The river breaks through the reef into an alluvial valley in the Blue Hills and Blue Valley on the Mancos Formation. The Fremont joins Muddy Creek near Hanksville and becomes the Dirty Devil River.

Before about 1890, the Fremont River near Hanksville was a narrow, meandering river about 6.5 feet deep and 100 to 328 feet wide. Several large floods in a twenty-year period from 1890 to 1910 eroded about 15 feet into the floodplain and produced a braided river 19 to 23 feet deep and up to 1,000 feet wide (Graf 1980; Everitt and Godfrey 2005). Since then, vegetation has grown on the river banks, and the channel is returning to the narrow, meandering character.

The upper reaches of the Fremont River contain brown and rainbow trout down to about Torrey, Utah. Whirling disease has taken its toll and reduced the local fish population (Cook 2000). The Fremont River from Johnson Valley Reservoir to Mill Meadow Reservoir contains rainbow and brown trout. From Mill Meadow to Torrey, the Fremont River is dewatered for irrigation at times, but brown trout are still present. The best fishing below Mill Meadow Reservoir is at Bicknell Bottoms on Division of Wildlife Resources' Bullock Waterfowl Management Area (Prettyman 2001). From Torrey to the confluence with Muddy Creek, the Fremont River hosts only warm-water fish species.

Dirty Devil River

This river was named by the Powell expedition. On July 28, 1869, one of the boats turned from the Colorado River into this tributary, and in response to a question about trout from the following boat, William Dunn responds that it is a "dirty devil" (Powell 1957). The Dirty Devil with its numerous braided channels flows southwest and begins to meander west and south as it flows to the confluence with the Colorado River in Lake Powell (Map 56). The Dirty Devil River cuts back down-section through Morrison, Curtis, Summerville, Entrada, Carmel, and Navajo formations as it meanders into the west flank of the Monument Upwarp. The Dirty Devil swings in a wide arc around the Henry Mountains, cuts a deep, steep-walled canyon into

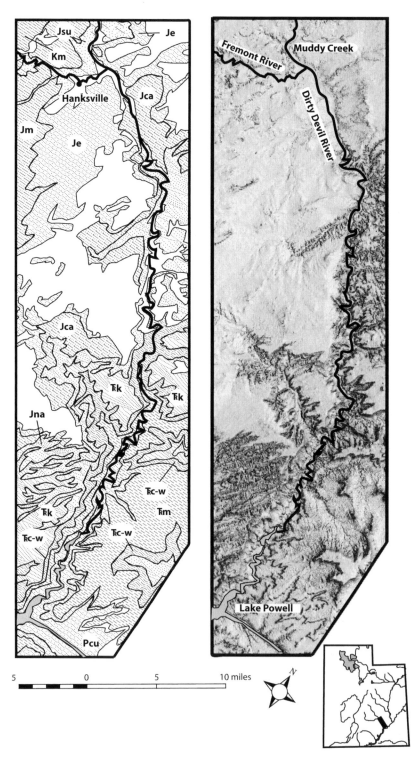

Map 56. Shaded relief and geologic maps of the Dirty Devil River. Formation abbreviations are: Km = Mancos Shale, Jm = Morrison Formation, Jsu = Summerville Formation, Jc = Curtis Formation, Je = Entrada Formation, Jca = Carmel Formation, Jna = Navajo Sandstone, Ŧk = Kayenta Formation, Ŧc-w = Chinle and Wingate formations, Ŧm = Moenkopi Formation, Pcu = Cutler Group.

Figure 24. Canyon view up the Dirty Devil River from the mouth of Hatch Canyon. Two small alcove arches can be seen in the Wingate Sandstone cliff. The Chinle Formation forms the slopes and Moenkopi Formation is at river level. U.S. Geological Survey photograph by R. L. Miller, Garfield County, Utah, 1939, Figure 8-B in Hunt (1953).

Kayenta and Wingate Sandstone (Figure 24). The river continues into the Chinle, Moenkopi, Permian White Rim Sandstone, and Cedar Mesa Sandstone.

San Rafael River

The San Rafael River, named by early Spanish Franciscan friars after the Archangel Saint Rafael (Van Cott 1990), begins in Castle Valley where Huntington Creek, Cottonwood Creek, and Ferron Creek join in west-dipping Dakota Sandstone and Morrison Formation. The upper part of Huntington Creek was diverted into the San Rafael River by capture near Lawrence. Huntington Creek originally flowed down Buckhorn Draw (Stokes 1978). The San Rafael River flows against the dip of the west-dipping Morrison, Summerville, Curtis, Entrada, and Carmel formations into the heart of the San Rafael Swell. The river meanders into steep-walled canyons in the Navajo, Kayenta, Wingate, Chinle, and Moenkopi formations. Navajo Sandstone caps the Wedge above on the north, and Wingate Sandstone caps Assembly Hall Peak on the south. Navajo Sandstone caps Window Blind Peak south of Assembly Hall. The river turns from east to southeast and cuts into Permian Kaibab and the Cedar Mesa Sandstone Member of the Cutler Group in the core of the San Rafael Swell. Meandering on its course all the way, the river then crosses the San Rafael Reef in steeply eastward-dipping Wingate, Kayenta, Navajo, and Carmel formations and meanders

175

Map 57. Shaded relief and geologic maps of the Price Canyon section of the Price River. Formation abbreviations are: Tg = Green River Formation, Tco = Colton Formation, Tf = Flagstaff
Formation, Kpr = Price River Formation, Kbh = Blackhawk Formation, Km = Mancos Shale.

southeast across the San Rafael Desert. Then the river curves northeast and joins the Green River.

Price River

On topographic maps, the Price River begins at the Scofield Reservoir outlet, but local fishermen refer to the section from Scofield Reservoir to the confluence with the White River as lower Fish Creek and the Price River begins where the White River joins. The river flows in a steep canyon on the high Wasatch Plateau east and north to the White River confluence near Colton (Map 57). The former course of the Price River followed a strike valley in the Lower Colton Formation, into Minnie Maud Creek, and then to Nine Mile Canyon and the Green River. This upper Price River was captured due to headward erosion of the river that occupied Price Canyon. About 25 miles of the strike valley in the Colton Formation were abandoned (Stokes 1978).

The Price River flows eastward into Price Canyon and cuts down through the Tertiary lake sediments, Flagstaff, Colton, and Green River formations of the Roan Cliffs exposed to the north, flows down into the Cretaceous sediments of the Book Cliffs, the Price River, Castlegate Sandstone, and Blackhawk Formation to Helper, then into various members of the Mancos Shale south and southeast to Price. The river meanders southeast in slope wash and Mancos Shale to Wellington, then cuts lower into the Mancos and crosses the Farnham anticline flowing south into Dakota Sandstone. The Price River then progresses southeast across the north end of the San Rafael Swell through the Morrison Formation (Map 58). The river meanders deeper into Summerville, Curtis, and Carmel formations, then back up-section in reverse order into Dakota Sandstone and Mancos Shale to Woodside, Utah. The river then plunges into a steep canyon in the Book Cliffs, meandering in a steep-walled canyon. The gorge is in Blackhawk Formation at river level, with Castlegate, Bluegate, and Price River formations above, but joins the Green River in upper Mancos Shale.

The Price River from Scofield Dam to the confluence with the White River, known as Fish Creek, contains dominant brown trout with some rainbow and cutthroat trout and is stocked annually (Cook 2000; Demoux 2001). This stretch of river is either high and muddy or dewatered in the fall when the gates of Scofield Dam are closed. Fish Creek above Scofield Reservoir has 10 miles of quality fly-fishing for brown, rainbow, and cutthroat trout (Prettyman 2001).

Map 58. Shaded relief and geologic maps of the lower Price River to the confluence with the Green River. Formation abbreviations are:
TKnh = North Horn Formation,
Kpr = Price River Formation,
Kbh = Blackhawk Formation,
Km = Mancos Shale,
Kdcm = Dakota Sandstone and Cedar Mountain Formation,
Jm = Morrison Formation,
Jscu = Summerville and Curtis formations,
Je = Entrada Formation,
Jca = Carmel Formation.

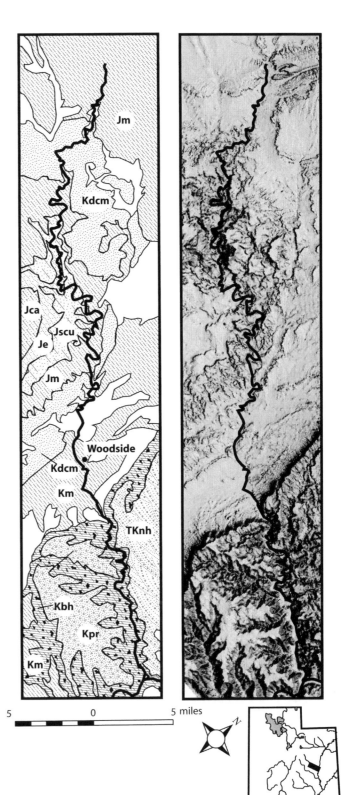

5   0   5 miles

The Fish

An interest in rivers might be accompanied by an interest in the freshwater fish that inhabited the rivers. Fossil fish provide a means of tracking changes in river systems in the Late Cenozoic, because fish quickly colonize suitable habitats when new river connections are made, and changes in river systems happen too quickly for the fish to evolve adaptations (Smith 1999). The distribution of similar fish species in two separate, but adjacent river systems therefore implies a previous connection, and evolutionary divergence of geographically isolated species implies a barrier that lasted for millions of years. For example, fossil fish suggest to Smith (1999) that the upper Snake River tributaries were connected to the Virgin and Lower Colorado rivers, and that the lower and upper Colorado were separate in the Miocene. Fish records also indicate low elevations in the northern Utah–southern Idaho region 6 million years ago.

Some fish are closer to humans than to some other fish. In fact, humans and fish share many common features including cranium, brain, eyes, ears, nose, dermal skeleton, internal bony skeleton, vertebral column with notochord, thyroid, pituitary gland, liver, gall bladder, pancreas, and spleen, all early adaptations to an aquatic environment. Trout are closer cousins to humans than they are to sharks, and coelacanths are even closer cousins to humans than are trout (Dawkins 2004).

Fish are spectacularly diverse with about 25,000 living species. One new fish species appeared about every 18,000 years. Compare fish evolution with the evolution of birds. The oldest bird is archaeopteryx that appeared 150 million years ago. Today there are about 10,000 living bird species.

Fish ancestry began long before dinosaurs (Gilbert 1992; Maisey 2000; Dawkins 2004). The first vertebrates were fish, the ostracoderms, which appeared in the Cambrian 510 million years ago, but these animals became extinct in the Devonian 350 million years ago. They were jawless fish similar to varieties of living jawless fish such as lampreys and hagfish. Jawless fish dominated seas in the Devonian age of fishes (Figure 25).

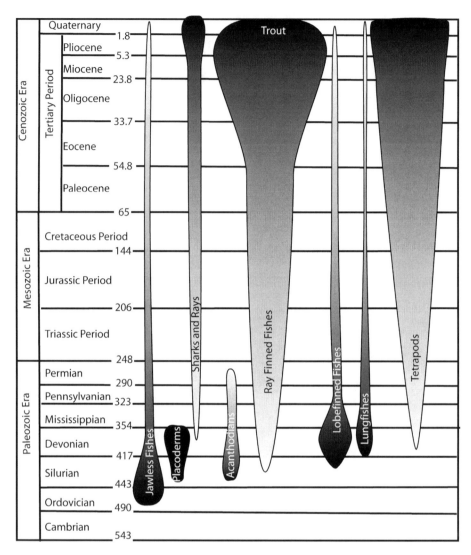

Figure 25. Evolution of fish through geologic time (modified from http://hoopermuseum .earthsci.carleton.ca/coelacanth/F4.htm, accessed Jan. 25, 2008.

Mammalian evolution is rooted in fish beginning 360 million years ago, when transitional animals with gills and limbs first appeared. The approximate timing of ancestral splits between lungfish, coelacanths (lobefins), mammals, and ray-finned fish begins at 440 million years ago when ray-finned fish split from the other groups. Ray fins have a skeleton like a Victorian fan. The ray fins lack a fleshy lobe at the fin base. Then at 425 million years ago coelacanths split off. At 415 to 420 million years ago, lungfish split off leaving tetrapods. The first jawed fish were the acanthodians that appeared in late Silurian (410 million years ago) and became extinct be-fore the end of the Permian (250 million years ago). The acanthodians were small, shark-like fish that may be related to modern bony fish (Osteich-thyes). Modern fish evolved from different ancestors than tetrapods.

Tetrapods have four limbs. Tetrapods are offshoots of bony fish (Sar-copterygii); other members are coelacanths and lungfish (superorder Dip-noi). Refer to Mcfarland and colleagues (1985) for details of classification of fish.

Evolution of bony fishes continued in ever-expanding fashion. The plac-oderms, another group of jawed fish, appeared at the beginning of the De-vonian 395 million years ago and became extinct at the end of the Devonian about 345 million years ago. They were small, flattened bottom dwellers.

There are two large classes of higher fishes, the sharks and other cartilag-inous fish (Condrichthyes) and the bony fishes (Osteichthyes). Both classes first appeared in the Devonian. Cartilaginous-skeleton fish, the sharks and rays (class Chondrichthyes), appeared 370 million years ago in middle De-vonian (Colbert 1969). Modern bony fish (class Osteichthyes) appeared in late Silurian or early Devonian about 395 million years ago. The early forms were found in fresh water. There are no marine fossils older than Triassic, about 230 million years ago. The ray-finned fish (Actinopterygii), subclass Osteichthyes, became the dominant fish. The ancestors of land vertebrates are bony fish (Choanichthyes or Sarcopterygii), the lobe-finned (superorder Crossopterygii) fish that appeared in Silurian or early Devonian at 390 mil-lion years ago (Figure 25). Humans are lobefins adjusted for life on land. The group of lobefins from which we tetrapods are derived are today a few pitiful genera, but once dominated the seas.

Ray-finned fish (subclass Actinopterygii), one of two major divisions of true bony fish (osteichthyans), were distinct as early as Devonian, and are the most abundant and diverse group of backboned fish on earth today, with 23,681 species and 42 orders. The three groups of modern ray-finned fish are first, the primitive reed fish; second, cartilage-bone paddle fish and stur-geon (infraclass chondrosteans); and third, neopterygians, which include gars, bowfins, and teleosts. Teleost fish are whole-boned fish that evolved into a myriad of different kinds of fish. Earliest fossil teleosts are found in Middle Triassic rocks 235 million years old. Subgroups of teleosts in-clude some more advanced fish including pikes, mudminnows from the late Cretaceous in Alberta, and the Salmoniforms—salmon, trout, and smelts—found in the early Cretaceous. By the Eocene, smelts and salmonids were separate lineages. By about 15 million years ago, common ancestral trout split into two branches. One branch gave rise to Pacific salmon, rainbow, and cutthroat trout. The other branch gave rise to Atlantic salmon. Many lines of bony fish evolved through Mesozoic. No animal that lived in water has been as successful as the main line of bony fishes (Actinopterygians).

Modern Fish

Freshwater rivers and lakes comprising about 0.01 percent of water on earth are the habitat for 40 to 45 percent of fish species. Much of the 2.5 percent of Earth's water that is fresh is locked in ice caps or groundwater, leaving little open surface water to support the enormous diversity of freshwater fish.

Geology and hydrology exert first-order controls on salmonid fish populations, and geologic history of basins controls the hydrologic, sediment transport, and temperature that establish the environmental niches to which fish have adapted. These factors also established their migration routes. The geographic and geologic environment for fish has changed dramatically in Tertiary time. The fish species are slower to change and the species have been present longer than the geographic details of the river systems in which the fish occur (Smith 1981).

The Platte River, a part of the Mississippi River system, into which the ancestral Green River flowed, includes many fish families adapted to seasonally warm water. A few cold water salmonids (trout, whitefish) are present in mountain headwaters in areas of geologically recent transfers across the Continental Divide from the Columbia River basin. The North Platte has no salmonids (Hansen 1986). The Upper Colorado and Green have narrowly limited fish species with only four families and fourteen species that live in low-altitude, seasonally warm water and high-altitude cold water. Warm-water species, endemic and restricted to the Colorado River, have evolved adaptations unique to the Colorado. Cutthroat trout and whitefish, cold-water nonendemic species, were transferred from the Columbia River in recent geological time. Neither of these species was endemic to the Platte River.

Salmonids are a cold-water, holarctic group including some species that enter fresh water only to breed. Salmoninae (salmon and trout), Coregoninae (whitefish), and Thymallinae (graylings) are the three subfamilies. Salmonids have been in North America since the Eocene. Trout belong to the perch and trout family, Salmonidae.

Brook trout belong to the genus Salvelinus, species *S. fontinalis*. Lake trout is *S. namaycush*. Rainbow trout belong to genus Salmo, species *S. gairdnerii*. Cutthroat is *Oncorhynchus clarkii* and brown trout is *S. trutta*. Brook trout, *Salvelinus fontinalis*, is native to northeastern North America from Georgia to the Arctic Circle and were introduced to other areas of the United States, Canada, South America, and Europe. Brown trout, *Salmo trutta*, is a native of Europe from the Mediterranean basin to the Black Sea to Arctic Norway and Siberia, and were widely introduced to North America beginning in 1883. Rainbow trout, *Salmo gairdneri*, is a native American trout, and its natural range includes the mountains of northern Mexico to the Aleutian Islands. Golden trout, *Salmo aguabonita*, is native to headwaters of the Kern River in California. Cutthroat trout, *Salmo clarkii*, is endemic to the western United States and is also sometimes called native trout (McClane 1965). Troutperch, *Percopsis omiscomaycus*, is neither trout nor true perch. It has an adipose fin on its back and spiny rays on dorsal and anal fins. Troutperch is found naturally in the Yukon, Hudson Bay, Lake Champlain, Potomoc River, Great Lakes, south to West Virginia, Kentucky, and Kansas.

Tertiary Fish

In Tertiary time, the Uinta Mountains formed the division between two lakes. Fish in the lakes—gar, catfish, perch, bowfins, paddlefish, herrings, and sunfish—are all members of eastern families, suggesting that the lakes drained east into the Mississippi River system. None of the fourteen species of fish native to the present upper Colorado River had progenitors in the Green River lakes. Indigenous Salmonidae (trout) of the modern upper Green River and Columbia are known in North America from the Eocene, but are not present in the Green River Formation (Eocene age) (Miller 1965; Patterson 1981).

Colorado squawfish and other fishes of the Colorado River system have been geographically isolated for millions of years. They have adapted to fast, seasonally warm, turbid waters of the Green River (Hansen 1996).

The Tertiary lakes north and south of the Uinta Mountains contained a rich fish fauna, but the extent to which these fish migrated into the mountain rivers feeding the lakes from the surrounding mountains is not known. The lakes were in a subtropical to temperate climate, but the rivers originated in the high-altitude mountains. One of the most abundant of the fish inhabitants of the lakes was *Knightia alta*, a small herring-like fish usually about 8 inches long, but some 20-inch trophy fossils have been found. *Diplomystus dentatus* (Figure 26) was a surface feeder and also consumed Knightia. Large gars and troutperch were also found in the lakes. Troutperch are commonly called trout, but belong to a different order and family than trout. Troutperch belong to the order Percopsiforms, family Percopsidae; trout belong to the order Salmoniforms, family Salmonidae. Two rare

Figure 26. Fossil Eocene-age fish *Diplomystus dentatus* showing bones, fin ray, and other features, embedded in a slab of rock taken from the quarry at Fish Cut. Fossil found in the Green River Formation, Sweetwater County, Wyoming. U.S. Geological Survey photograph by W. T. Lee, 1916, Plate XIX in U.S. Geological Survey Bulletin 612.

trout from the Green River formation, Erismatopterus and Amphiplaga, aren't really in the trout family at all, but the troutperch family. The Green River Formation also includes fly and mosquito fossils, insects that could have been food for surface-feeding fish.

Trout did not appear until the Tertiary lakes were gone. Trout first appeared in the Miocene. Salmon (Salmonidae) were present from the Eocene, brook trout (Salvelinus) were present in the Pleistocene from the Yarmouth interglacial, and Prosopium were present in the Wisconsin glacial period (Miller 1965).

The dominant families of freshwater fishes in the late Cenozoic (Pliocene to Recent) in North America according to Smith (1981) are gars (Lepisosteidae), trouts and salmon (Salmonidae), pikes (Esocidae), minnows (Cyprinidae), suckers (Catostomidae), catfishes (Ictaluridae), pupfishes (Cyprinodontidae), perch (Percidae), sunfish and bass (Centrarchidae), sculpins (Cottidae), and sticklebacks (Gasterosteidae). Barriers and long-term stability of aquatic habitat are the most important factors controlling species density and broad patterns of evolution. Because changes in species are much slower than changes in geography and climate, we cannot assume that fishes have recently adapted to the environments and climates in which they presently live.

The fish fauna of the West has an unstable history marked by many extinctions. The fauna of the West has only about one-fourth as many species as the East. Mountains have existed long enough to allow trout, whitefish, mountain suckers, dace, and sculpins to adapt to mountain rivers. These species adapted to headwaters on the major barriers and cross them frequently. Western freshwater habitats are subdivided and isolated by mountain ranges into river basins and several hundred interior drainages. The Miocene and Pliocene Salt Lake beds in northwest Utah show faunal similarities with the Snake River Plain. Fish of Cache Valley Formation in northeastern Utah are different, indicating separation by a drainage divide in south-central Idaho.

Miocene fish fauna in the Bear River and upper Snake River are distinctive relicts of northern and eastern forms such as whitefishes, suckers, and minnows. The course of the Bear River and easternmost upper Snake River in the Miocene may have connected to the northeast Hudson Bay drainage.

Cretaceous Fish

The Cretaceous rivers that flowed eastward from the high mountains in Utah and adjacent states into the sea must have harbored fish. Spiny-rayed fish (acanthomorph) are the largest and most advanced group of living ray-finned fish, and their oldest fossils appeared 160 million years or so ago. The remarkably successful infraclass teleost group of bony fishes began their expansion in the Cretaceous that continues through the Present (Colbert 1969). The first acanthomorph teleost fish were found in Cretaceous rocks in Utah (Stewart 1996).

Jurassic Fish

The Morrison Formation, deposited by the Morrison river system of Jurassic age, contains fairly diverse fish fossils. Four dipnoan (lung fish) taxa, five actinopterygians (ray-finned fish), and one teleost are described by Kirkland (1998). Fish are not rare in the Morrison, and permanent fresh water was abundant enough to occasionally support fish.

Triassic Fish

Fish were abundant in the Triassic and their fossils are found in the Chinle Formation as well as in the correlative units such as the Dockum of Texas (Huber, Lucas, and Hunt 1993). Primitive bony fishes (palaeoniscoid) were common. The first bony fish (Chondrosteans) were replaced by holosteans in Triassic and Jurassic, and in Cretaceous to Cenozoic they were replaced by teleosts. Ray-finned fish (Chondrichthyans, Actinopterygians) were common in Mesozoic marine and fluvial environments. Ray-finned fish (subclass Osteichtyes) became the dominant fish.

Lungfish (Arganodus) were also common in freshwater during the Triassic and Jurassic, possibly because the climate was monsoonal at the time, but disappear in the Cretaceous. Coelacanths, lobe-finned fish related to more modern but no longer extant coelacanth (Latemaria), were common marine and freshwater species but disappeared after the middle Jurassic (Wilson and Bruner 2004). The coelacanths are widely reported from the Triassic of the western United States including the Chinle Formation. The Chinle rivers hosted freshwater sharks (Xenacanthus and Lissodus). The Springdale Sandstone, which is near Chinle Formation in age, has large, sturgeon-like, freshwater fish.

Atmospheric oxygen levels play a key role in the evolution of metazoans and may have played a role in the evolution of fish. The carbon and sulfur geochemical cycles control atmospheric oxygen. Atmospheric oxygen concentration varied throughout the Phanerozoic, with a maximum at 300 million and a minimum at 200 million years ago, and an overall increase from 200 million years to present. Isotopic record of organic carbon over the past 205 million years suggests oxygen levels doubled from 10 volume percent to 21 volume percent (the present-day value) over this period (Falkowski et al. 2005). Examination of the fossil record suggests atmospheric oxygen is potentially a key in evolution of large placental mammals in the Cenozoic. The relatively rapid decline in late Permian and early Triassic contributed to extinction of terrestrial animals (mostly reptiles). Changes in atmospheric oxygen are driven by tectonics and increased burial efficiency of organic matter on continental margins (Falkowski et al. 2005; Berner, Vanden Brooks, and Ward 2007).

Conclusion

Utah is a treasure house of geology, resources, scenery, and recreation. Many of the scenic treasures are the result of Utah's diverse river system. Bryce Canyon formed from headward erosion of tributaries of the Paria River. Zion Canyon was formed by the Virgin River system. Canyonlands is the result of the Green and Colorado rivers eroding the landscape. Recreational opportunities are abundant along Utah's rivers. Gary Nichols (2002) describes forty-seven rivers that can be run in a raft or kayak. James Demoux (2001) describes fifty-seven rivers that offer fly-fishing.

River erosion has exposed much of Utah geology for observation and study. The deep canyons of the Colorado and Green rivers have exposed some of the oldest rocks in Utah and the sequence of strata that covers them. River erosion has exposed rich mineral deposits of copper and uranium on Utah's Colorado Plateau for exploitation. River deposits record the advancing thrust sheets involved in the complex mountain-building history. Utah rivers record changes that have occurred in climate with changing sediment load and down-cutting that left abandoned river terraces. Entrenched meanders on the Colorado Plateau are sensitive indicators of recent uplift.

Utah is different from adjacent states. Nevada is all Basin and Range with no Colorado Plateau or Rocky Mountains; Idaho also has no Colorado Plateau; Wyoming has no Colorado Plateau or Basin and Range; Colorado has no Basin and Range and no Sevier-age mountains; Arizona is all Basin and Range and Colorado Plateau with no Middle Rocky Mountains. Utah's history of active mountain building that continues today has been responsible for many adjustments in river drainage. Utah has a varied and diverse river system that has examples of river piracy, antecedence, superposition, and drainage reversals.

Utah's ancient river systems flowed from mountain ranges that are now gone, and the rivers flowed through landscapes that have been obliterated by younger events. Mesozoic- to Tertiary-age mountain-building events produced rivers across the entire state that left their deposits across Utah.

Geologists use the river deposits of conglomerate to establish the sequence and timing of mountain building.

Rivers in Utah cross geological barriers with impunity. The Green River crosses the most formidable barrier, the Uinta Mountains. This crossing involves a complex sequence of geological events including mountain building and mountain collapse that caused a drastic reorganization of river drainage in the eastern Uinta Mountains. The Provo and Weber rivers cross the Wasatch Mountains. Blacksmith Fork and the Little Bear River cross the Bear River Range.

The Colorado River crosses barrier after barrier in its course through Utah to the Grand Canyon of Arizona. The barriers include faults, salt-cored anticlines, and uplifts. The Colorado River system has removed thousands of feet of sediment from the Colorado Plateau. No river in the Western Hemisphere has crossed more barriers or eroded more than the Colorado River.

The Bear River leaves Utah to enter Wyoming and Idaho, but then reverses its course and returns to flow into Great Salt Lake rather than into the Snake River system.

The Sevier River reverses an ancient course that formerly carried it to the vicinity of Utah Lake to finally empty into a nearly dry basin of Sevier Lake. In its course northward, the Sevier River has cut through two steep canyons at Circleville and Marysvale, probably as a result of superposition.

The Provo River greedily captured much of the upper Weber River drainage, but then carried the water into Utah Lake and finally into Great Salt Lake where the Weber also ends. A tributary of the Weber captured Chalk Creek and left a series of barbed Chalk Creek tributaries to tell the tale.

Utah's rivers should be treasured and enjoyed. They provide a fascinating geologic account, breathtaking scenery, a wealth of natural resources, and abundant recreational opportunities.

Eventually, all things merge into one,
and a river runs through it.

— NORMAN MACLEAN, *A River Runs Through It*

Glossary

Acanthodii (spiny sharks) A class of extinct fish having features of bony fish (Osteichthyes) and cartilaginous fish (Chondrichthyes). They appeared in the Early Silurian and disappeared in the late Permian. They were the first known jawed vertebrates.

acanthomorph Spiny-rayed fish. The largest and most advanced group of living ray-finned fish.

accreted terrane A crustal block with a distinctive geologic history different from surrounding areas that becomes attached to a continent.

Actinopterygii The ray-finned fish. The ray refers to the skeleton of their fins, which is similar to a Victorian lady's fan.

aggradation Raise the level of the bottom of a river by deposition of sediment.

aggrading system A river system that deposits sediment raising the river level.

alluvial fan A fan-shaped deposit of sand and gravel formed by a river as it flows from a mountain range into a valley.

alluvial plain Sediment consisting of gravel, sand, and silt accumulated on the floodplain of a river.

alluvial river A river flowing over sediment deposited by a river.

alluviated A surface formed by river erosion.

annular river A river channel in the form of a ring around some geological obstacle.

antecedent river A river that is older than structures that it crosses.

antedates Predates.

anticlinal axis The axis of a fold in which the layers of sediment on either side are inclined away from the crest.

anticlinal uplift An uplift or fold in which the layers of sediment on either side are inclined away from the crest.

ash beds Layers of finely divided fragments of volcanic rock.

back-bulge basin A shallow, broad zone of flexural subsidence toward the continent from a forebulge high (see forebulge high).

barbed tributary A tributary that enters another river in an upriver direction.

basalt A volcanic rock rich in iron giving the rock a dark color. The rock is a fine-grained assemblage of igneous minerals.

base level The lowest level to which a river can erode its bed. Lakes and the ocean form base levels for rivers.

Basin and Range extension The Basin and Range is a large physiographic province in the western United States that is characterized by north-south-trending mountain ranges and intervening basins. The boundary between mountain and basin is a normal fault that results from extension.

bedrock river A river channel cut into bedrock.

blind valley A valley containing a stream that disappears into the subsurface before exiting the valley.

block faulting Normal faults that break segments of the crust into a series of upthrown and downthrown blocks.

bowfins An order of primitive ray-finned fish. Six families are known from Jurassic, Cretaceous and Eocene rocks.

braided river A river with more sediment available than can be removed. The result is a complex network of channels separated by islands of sediment.

capture Occurs when a river from a neighboring drainage system erodes through the drainage divide between two rivers and captures another river, which then is diverted from its former course and flows down the course of the capturing river.

carbonates Rocks consisting of the carbonate minerals calcite and dolomite.

Choanichthyes or Sarcopterygii Bony fish ancestors of land vertebrates.

chondrosteans Cartilage bone fish such as the sturgeon and paddlefish.

clastic Rocks consisting of fragments of minerals and rocks that resulted from mechanical disintegration of the rock.

Coelacanths ("hollow spine") An order of the oldest living lineage of jawed fish known. Closely related to lungfishes.

competence The maximum size of mineral or rock fragments that can be moved by running water.

Condrichthyes Cartilaginous-skeleton fish—sharks and rays which appeared in Middle Devonian, 370 million years ago.

conglomerate A sedimentary rock composed of pebbles, cobbles, boulders, and sand.

consequent river A river whose course is determined by newly uplifted or emergent land surface.

continental shelf The margin of a continent that is beneath sea level.

convergence rate Rate of convergence of two adjacent lithospheric plates.

convergent lithospheric plate boundary Boundary between two adjacent lithospheric plates that are converging.

cosmic rays Protons and alpha particles (helium nucleii), electrons, gamma rays, and neutrons that originate beyond the Earth and strike the Earth's atmosphere.

Crossopterygii Superorder that includes the lobe-finned fish.

crust The outer layer of the Earth.

crystalline basement Older crystalline rocks covered by younger sedimentary rocks.

crystalline rocks Rocks composed of interlocking crystals such as igneous or metamorphic rocks. Crystalline rocks are often older rocks.

cut bank The outside of a meander bend.

debris flow Rocks and soil usually saturated with water that move rapidly down slopes.

delta Sediment deposited by a river at its mouth where it enters a quiet body of water such as a lake or an ocean.

dendritic river A group of river tributaries that form a branching pattern similar to the branches of some trees. Usually develops on rocks with uniform resistance to erosion or on flat-lying sedimentary rocks.

denudation Erosion of rocks exposing underlying rock and structure.

dermal skeleton Skeleton of vertebrates including fish and land mammals.

diapir An intrusion or piercing structure caused by buoyancy commonly applied to salt domes.

dip The angle of inclination measured with respect to a horizontal plane of a sedimentary bed or other planar geological feature such as a fault.

dip slip Slip on a fault in the direction of the dip of the fault.

Dipnoi Superorder of fish that includes coelacanths and lungfish.

discharge The volume of water in a river that moves past a given point in a unit of time.

discordant At a variance with the normal trend.

dolomite A mineral composed of calcium, magnesium, and carbonate.

dome An uplift of sedimentary rocks that is circular or elliptical in shape with the sedimentary beds dipping away from the center.

drainage divide Topographic divide separating two adjacent rivers.

endemic species A species whose natural occurrence is confined to a certain region and whose distribution is relatively limited.

entrenched meanders Meanders in a river developed on a floodplain that are cut into underlying rock.

extensional faulting Faulting that results from forces that extend the terrane such as normal faults.

fan delta An alluvial fan formed by river sediment accumulated at the base of a mountain range that transitions to an accumulation in a standing body of water such as a lake or an ocean.

Farallon plate A tectonic lithospheric plate that occurred in the Pacific Ocean of the western coast of the United States that was progressively consumed during convergence.

felsic dikes Igneous dikes (tabular bodies of igneous rock) that are composed of light-colored minerals rich in silicon.

floodplain The flat area bordering a river covered with sediment deposited during flood stages of the river.

fluvial gravels Gravels deposited by a river.

forebulge high The forebulge high (see foreland basin, back-bulge basin, and foredeep basin) is a broad region of flexural uplift related to lithospheric plate convergence.

foredeep basin A trough-like basin bordering a mountain belt between the structural front of the thrust belt and an upward flex known as the forebulge high (see foreland basin).

foreland basin Not to be confused with fore-arc basin, which is a basin parallel to a deep-sea trench and separated from it by a fore-arc ridge. A foreland basin system is an elongate region of crustal flexure that forms on continental crust between a thrust mountain belt and the adjacent stable continent that forms in response to processes related to lithospheric plate convergence. The system includes a foredeep basin between the structural front of the thrust belt and the upward flex known as the forebulge high (see forebulge high).

foreland bulge A typical mountain belt consists of, from west to east, a thrust belt, a foredeep basin, a forebulge high or foreland bulge, and a back-bulge basin.

glacial outwash Debris deposited by melt water from a glacier.

glaciers An accumulation of ice formed from compacted and recrystallized snow that is thick enough to flow downslope.

gneiss A metamorphic rock formed of alternating layers of coarse-grained, light-colored and dark-colored minerals.

graben A downthrown fault block that has been lowered in relation to fault blocks on either side.

gradient The slope of a river channel.

Great Basin A physiographic area of the western United States between the Rocky Mountains and the Sierra Nevada Mountains with interior drainage.

gypsum A mineral composed of calcium, sulfate, and water.

Hagfish Eel-shaped, jawless fish that resemble lampreys.

headward erosion A river lengthens its valley by eroding headward up the slope of the river.

holarctic A geographic division referring to the entire arctic region.

hydrometeorological Branch of meteorology that includes the hydrological cycle, water budget, and rainfall statistics of storms.

incision A river cutting into underlying rocks.

infraclass A rarely used further subdivision of a subclass in zoology.

intermediate dikes Igneous dikes (tabular igneous bodies) consisting of igneous rocks made of minerals that have an intermediate composition of silicon, iron, and magnesium.

intermontane Situated between mountains.

island arc A chain of volcanic islands that form an arcuate shape.

isotope A form of a chemical element that has a different number of nuclear neutrons but the same number of nuclear protons as other forms of the element.

karst Topography formed by groundwater dissolving underlying soluble rocks such as carbonates producing sink holes, caves, and solution valleys.

lacustrine Lake deposits.

lamprey A jawless fish with a toothed, funnel-shaped sucking mouth.

limestone A sedimentary rock composed primarily of the mineral calcite (calcium carbonate).

lithosphere The relatively rigid and strong outer layer of the earth that includes the crust and the upper, rigid mantle.

lobefin fish Superorder Crossopterygii appeared in Silurian or Early Devonian, 390 million years ago. Surviving members are lungfish and coelacanths. A fleshy lobe at the base of each fin gives the fish the name.

longshore currents A current in the shallow ocean that moves parallel to the shoreline.

lungfish Sarcopterygian fish belonging to the order Dipnoi. They are best known for their ability to breathe air and the presence of lobed fins with well-developed internal skeleton.

marine sediment Sediment deposited in an ocean environment.

mass wasting Movement of rock and soil downslope under the influence of gravity without involvement of running water.

meandering river A river with a series of broad, looping bends.

metamorphic rocks Rocks formed by recrystallization of preexisting rocks by application of heat, pressure, and chemically reactive fluids such as water.

monocline A single bend in horizontally layered rocks.

monsoon A wind pattern that changes direction depending on the season.

Monument Upwarp One of several Laramide age uplifts on the Colorado Plateau. Bounded on the east by a steep monocline, Comb Ridge.

moraine An accumulation of sand, gravel, and boulders deposited by a glacier.

Neopterygii ("new wings") A group of Actinopteri fish that appeared in the Late Permian. They have lighter scales and skeletons and more powerful jaws. Includes gars, bowfins, and teleosts.

neutrons A particle in the nucleus of an atom with no charge and a mass similar to the mass of a proton.

normal fault An inclined fault in which the side facing away from the inclined fault plane (hanging wall) has moved downward.

notochord The flexible cord of cells that forms the supporting structure of the body of all chordates, vertebrates, and several closely related invertebrates.

Oblique convergent lithospheric plate margin The margin of a lithospheric plate that converges with an adjacent plate at an oblique angle (not perpendicular or parallel).

obsequent river A river that flows in a direction opposite to the dip of the beds.

offshore bars A linear accumulation of sand related to a beach that forms just offshore.

orogeny A major episode of mountain building.

Osteichthyes A superclass of fish also called bony fish that includes the ray-finned fish (Actinopterygii) and lobe-finned fish (Sarcopterygii).

Ostracoderms ("shell skinned") Several groups of extinct, primitive, jawless fish that were covered in an armor of bony plates. Their fossils are found in Ordovician- and Devonian-age rocks in North America and Europe.

Pacific Ocean lithosphere The lithosphere consisting of crust and upper mantle rocks that underlies the Pacific Ocean.

parallel drainage pattern A river pattern developed on a sloping surface of homogenous rock.

passive continental margin The margin of a continent that is not involved in lithospheric plate convergence or divergence. The continental margin is not a plate margin.

physiographic province A province characterized by similar surface land-forms.

piracy or capture Occurs when a river from a neighboring drainage system erodes through the drainage divide between two rivers and captures another river, which then is diverted from its former course and flows down the course of the capturing river.

Placodermi Armored prehistoric fishes known from fossils dating from the late Silurian to the end of the Devonian period. The head and thorax were covered by articulated armored plates.

plate tectonics A theory of geology that has been developed to explain the observed evidence for large-scale motions of sections (plates) of the Earth's crust.

plugs An igneous intrusive body that fills the vent of a volcano.

plunging anticline An anticline with its axis inclined.

point bar A deposit of sand and gravel on the inside of a meander bend.

potash Salts of potassium such as potassium chloride.

quartzite A sandstone that has been metamorphosed or tightly cemented by silica cement.

radial river A river pattern that develops on a topographic high such as a dome or volcano where the rivers flow away from the highest elevation.

ray-finned fish Actinopterygii, the dominant group of vertebrates with 27,000 species ubiquitous throughout freshwater and marine environments.

rectangular drainage pattern A pattern of river channels with near right-angle bends that forms as the river follows joints, faults, and fractures that intersect.

resequent river An evolved consequent river such as a consequent river that initially flows down the slope of a newly formed anticline, but with time erodes into the anticline and follows a course determined by the most easily eroded beds.

river terrace A nearly level or gently sloping surface bordering a steeper slope along a river.

salmoniform Salmonidae A family of ray-finned fish, Salmoniformes, which includes salmon and trout. Atlantic salmon and trout are of the genus Salmo.

sand bars Linear accumulations of sand in a river or an ocean.

sandstone A sedimentary rock composed of sand-size (.0625 to 2 mm) particles cemented together.

Sarcopterygii or Choanichthyes Bony fish ancestors of land vertebrates.

schist A coarse-grained metamorphic rock composed of platy or elongate minerals such as mica.

shale A sedimentary rock composed of consolidated clay and mud.

shear stress (in river flow) The tangential stress exerted by flowing water against the river bed.

strike The direction measured with a compass of a horizontal line on the surface of a planar feature such as a sedimentary bed or fault.

strike-slip faulting A fault in which the movement direction is parallel to the strike of the fault.

strike valley A valley that parallels the strike of sedimentary layers.

subducting slab Slab of lithosphere at the edge of a plate that descends into the mantle.

subsequent river A river whose course is determined by the pattern of exposure of easily eroded rocks.

supercontinent A continental landmass that is an assemblage of more than one continental core.

superposed river or superimposed river A river that cuts downward through overlying rocks placing its channel on underlying rocks of different types or with different structure.

syncline A fold in layered rocks in which the sides of the fold are inclined toward the axis or center of the fold.

tectonic Features that are formed by major mountain-building events.

Teleostei One of three infraclasses in class Actinopterygii, the ray-finned fish. The other two infraclasses are Holostei and Chondrostei.

terrane A crustal block with a distinctive geologic history different from surrounding areas. Compare with "terrain," which refers to a tract of land.

tetrapod Four-limbed animal.

thrust fault A fault in which the side facing away from the inclined fault surface (hanging wall) has moved upward.

thrust plate Generally, the upper plate of a thrust fault.

thrust sheet Generally, the upper plate of a thrust fault.

tidalflats or mudflats Coastal wetlands that form when mud is deposited by the tides. They are formed in sheltered bays or estuaries.

Toreva block A fault block that forms when a strong sandstone or limestone overlies weaker rock such as shale or salt. The strong rock is broken into blocks that slide on the weaker material toward a cliff or riverbed. The blocks typically exhibit a backward rotation toward the parent cliff as sliding proceeds.

transverse canyon A river canyon that cuts across structures.

trellised river A river pattern that forms when resistant ridges of folded rocks force rivers to follow the easily eroded rocks between ridges, but at times the rivers cross the ridges.

tuff A fine-grained rock composed of fragments of volcanic rock (ash).

unconformity A gap in the succession of rocks that represents a buried erosion surface.

underfit river A river that is too small to have eroded the valley in which it flows.

volcanic ash Dust-sized particles of volcanic rock erupted from a volcano.

volcanic lahar A mudflow consisting of volcanic debris and water that flows down the slope of a volcano.

whirling disease A parasitic infection caused by a microscopic parasite, *Myxobolus cerebralis*. The parasite grows in the head and spinal cartilage of infected fish and causes a characteristic swimming behavior.

wind gap A former river channel through a ridge that is now abandoned due to piracy.

References

Alter, J. Cecil. 1941. Father Escalante's map. *Utah Historical Quarterly* 9:64–72.

Anderson, G. E. 1915. River piracy of the Provo and Weber rivers, Utah. *American Journal of Science* 40: 314–316.

Anderson, John J. 1987. Late Cenozoic drainage history of the northern Markagunt Plateau, Utah. In *Cenozoic geology of western Utah: sites for precious metal and hydrogen accumulation,* ed. R. S. Kopp and R. E. Cohenour, 271–277. Utah Geological Association Publication 16.

Anderson, John J., and Peter D. Rowley. 1975. Cenozoic stratigraphy of southwestern high plateaus of Utah. In *Cenozoic Geology of southwestern high plateaus of Utah*, ed. J. J. Anderson, Peter D. Rowley, R. Fleck, J. Kobert, and A. E. M. Nairn, 1–51. Geological Society of America Special Paper 160.

Anderson, R. E., and Gary E. Christenson. 1989. Quaternary faults, folds, and selected volcanic features in the Cedar City 1 degree × 2 degree Quadrangle, Utah. Utah Geological Survey Miscellaneous Publications Report 89-6.

Andrews, Edmund D. 1997. Formation of the Zion Narrows slot canyon. Supplement, *Eos, Transactions, American Geophysical Union* 78, no. 46: 272.

Ash, Sidney. 2003. The Wolverine petrified forest. Utah Geological Survey, *Survey Notes* 35, no. 3: 3–6.

Auerbach, Herbert S. 1941. Old trails, old forts, old trappers, and old traders. *Utah Historical Quarterly* 9: 13–63.

Baars, Donald L., ed. 1973. *Geology of the canyons of the San Juan River*. Durango, Colo.: Four Corners Geological Society.

———. 1987. *Cataract Canyon via the Green or Colorado rivers: A river runner's guide*. Evergreen Colo.: Canon Publishers.

———. 1993. *Canyonlands country*. Salt Lake City: University of Utah Press.

———. 2000. *The Colorado Plateau: A geologic history*. Revised and updated. Albuquerque: University of New Mexico Press.

———. 2000. Geology of Canyonlands National Park, Utah. In *Geology of Utah's Parks and Monuments*, ed. D. A. Sprinkel, T. C. Chidsey, Jr., and P. B. Anderson, 61–83. UGA Publication 28.

Baars, Donald L., and Gene Stevenson. 1986. *San Juan Canyons: A river runner's guide and natural history of San Juan River Canyons*. Evergreen, Colo.: Canon Publishers.

Baars, D. L., and Gary C. Huber. 1973. River log. In *Geology of the canyons of the San Juan River*, ed. D. L. Baars, 55–94. Durango, Colo.: Four Corners Geological Society.

Bancroft, H. H. 1889. *History of Utah 1540–1886*. San Francisco: The History Company.

Belknap, Bill, and Buzz Belknap. 1974. *Canyonlands river guide: Westwater, Lake Powell, Canyonlands National Park*. Boulder City, Nev.: Westwater Books.

Belknap, B., and L. B. Evans. 1995. *Belknap's revised waterproof Dinosaur river guide*. Evergreen, Colo.: Westwater Books.

Berner, Robert A., John M. Vanden Brooks, and Peter D. Ward. 2007. Oxygen and evolution. *Science* 316: 557–558.

Bierman, Paul R. 1994. Using in situ produced cosmogenic isotopes to estimate rates of landscape evolution: A review from the geomorphic perspective. *Journal of Geophysical Research* 99: 13,885–13,896.

Billingsley, George H., Peter W. Huntoon, and William J. Breed. 1987. *Geologic map of the Capitol Reef National Park and vicinity, Emery, Garfield, Millard, and Wayne Counties*. Salt Lake City: Utah Geological Survey.

Bradley, Wilmot H. 1936. *Geomorphology of the north flank of the Uinta mountains*. U.S. Geological Survey Professional Paper 185-I.

Bruhn, R. L., M. D. Picard, and J. S. Isby. 1986. Tectonics and sedimentology of the Uinta Arch, Western Uinta Mountains, and Uinta Basin. In *AAPG Memoir 41*, 333–352.

Bryan, Kirk, and E. C. LaRue. 1927. Persistence of features in an arid landscape: The Navajo Twins, Utah. *Geographical Revue* 17, no. 2: 251–257.

Bryant, Bruce. 1990. *Geologic map of the Salt Lake City 30' × 60' quadrangle, north-central Utah, and Uinta County, Wyoming*. U.S. Geological Survey Map I-1944.

Bryant, Bruce, D. J. Nichols, and W. A. Cobban. 1982. *Late Cretaceous orogenic pulses in north central Utah*. U.S. Geological Survey Professional Paper 1375.

Carson, Eric C. 2005. Fluvial geomorphology and hydrology of the sub-alpine rivers of the Uinta Mountains. In *Uinta Mountain geology*, ed. Carol M. Dehler, Joel L. Pederson, Douglas A. Sprinkle, and Bart J. Kowallis, 171–187. Utah Geological Association Publication 33.

Cater, F. W. 1970. *Geology of the salt anticline region in southwestern Colorado*. U.S. Geological Survey Professional Paper 637.

Cline, Eric J., and John Bartley. 2001. *Combined extensional-detachment and salt tectonics form an atypical extensional basin, Sevier Valley, Utah*. Geological Society of America, Abstracts with Programs, v. 33, 391–392.

———. 2002a. *An upthrown valley? Unconventional faulting in Sevier Valley, Utah*. Geological Society of America, Abstracts with programs, v. 34, no. 4: 61.

———. 2002b. *Southern termination of the Wasatch fault localized by evaporate-cored Cretaceous anticline, Sevier Valley, Utah*. Geological Society of America, Abstracts with programs, v. 34, no. 6: 438.

Colbert, Edwin H. 1969. *Evolution of the vertebrates*. 2nd ed. New York: John Wiley.

Constenius, K. N. 1996. *Late Paleogene extensional collapse of the Cordilleran foreland fold and thrust belt*. Geological Society of America Bulletin 108: 20–39.

Cook, Steve. 2000. *Utah fishing guide*. Salt Lake City: Bryan Brandenburg, Utah Out doors.com.

Counts, Ron, and Joel L. Pederson. 2005. The nonglacial surficial geology of the Henrys Fork, Uinta Mountains, Utah and Wyoming. In *Uinta Mountain geology*, ed. Carol M. Dehler, Joel L. Pederson, Douglas A. Sprinkle, and Bart J. Kowallis, 155–169. Utah Geological Association Publication 33.

Craig, L. C., C. N. Holmes, R. A. Cadigan, V. L. Freeman, T. E. Mullens, and G. W.

Weir. 1955. *Stratigraphy of the Morrison and related formations, Colorado Plateau region: A preliminary report.* U.S. Geological Survey Bulletin 1009-E.

Davis, W. M. 1901. An excursion to the Grand Canyon of the Colorado. Harvard College Museum of Comparative Zoology 38: 107–201.

Dawkins, Richard. 2004. *The ancestors tale: A pilgrimage to the dawn of evolution.* New York: Houghton Mifflin Company.

Dawson, James Clifford. 1970. The sedimentology and stratigraphy of the Morrison formation (upper Jurassic) in northwestern Colorado and northeastern Utah. Ph.D. diss., University of Wisconsin, Madison.

DeCelles, Peter G. 1988. Lithologic provenance modeling applied to the Late Cretaceous synorogenic Echo Canyon Conglomerate, Utah: A case of multiple source areas. *Geology* 16: 1039–1043.

————. 1994. *Late Cretaceous-Paleocene synorogenic sedimentation and kinematic history of the Sevier thrust belt, northeast Utah and southwest Wyoming.* Geological Society of America Bulletin 106.

DeGrey, Laura D., and Carol M. Dehler. 2005. Stratigraphy and facies analyses of the eastern Uinta Mountain Group, Utah-Colorado border Region. In *Uinta Mountain geology,* ed. Carol M. Dehler, Joel L. Pederson, Douglas A. Sprinkle, and Bart J. Kowallis, 31–47. Utah Geological Association Publication 33.

Demoux, James B. 2001. *Flyfisher's guide to Utah.* Belgrade, Mont.: Wilderness Adventure Press.

Dickson, Timothy A., Justin M. Fairchild, Jennifer L. Coor, Christopher M. Bailey, and David W. Marchetti. 2006. *Grabens of the Fish Plateau, High Plateaus, Utah.* Geological Society of America, Abstracts with Programs, v. 38, no. 6: 11.

Doelling, Hellmut H. 1988. Geology of the Salt Valley anticline and Arches National Park, Grand County, Utah. In *Salt deformation in the Paradox region, Utah,* ed. H. H. Doelling, C. G. Oviatt, and P. W. Huntoon, 1–60. Utah Geological and Mineral Survey Bulletin 122.

————. 2001. *Geologic map of the Moab and eastern part of the San Rafael Desert 30' × 60' quadrangle, Grand and Emery Counties, Utah, and Mesa County, Colorado.* Utah Geological Survey Map 180.

Dougher, Frank L., and Michael P. O'Neill. 1997. *An evaluation of the geomorphic characteristics of a threshold meandering/anastomosing channel system.* Geological Society of America, Abstracts with Programs, v. 29, no. 6: 256.

Dover, J. H. 1995. *Geologic map of the Logan 30' × 60' quadrangle, Cache and Rich Counties, Utah, and Lincoln and Uinta Counties, Wyoming.* U.S. Geological Survey Map I-2210.

Dumitru, Trevor A., Ian R. Duddy, and Paul F. Green. 1994. Mesozoic-Cenozoic burial, uplift, and erosion history of the west-central Colorado Plateau. *Geology* 22: 499–502.

Dutton, C. E. 1880. *Geology of the high plateaus of Utah.* U.S. Geological Survey.

Eardley, Armand J., and Edward L. Beutner. 1934. Geomorphology of Marysvale Canyon and vicinity. Proceedings of the Utah Academy of Arts, Sciences, and Letters, v. 11: 149–159.

Emmons, S. F. 1897. The origin of the Green River. *Science,* new ser., 6: 19–21.

Enzel, Yehouda, Robert H. Webb, and Gerardo Benito. 1997. *Floods and arroyo formation of the east fork Virgin River, Utah.* Geological Society of America, Abstracts with Programs, v. 29, no. 6: 371.

Evans, Laura, and Buzz Belknap. 1973. *Flaming Gorge Dinosaur National Monument: Dinosaur River guide.* Boulder City, Nev.: Westwater Books.
———. 1974. *Desolation River guide: Green River wilderness.* Boulder City, Nev.: Westwater books.
Everitt, G. L., and A. E. Godfrey. 2005. *Channel shrinking and flood plain construction on the Fremont River near Hanksville, Utah.* Geological Society of America, Abstracts with Programs, v. 37, no. 6: 40.
Falkowski, Paul G., Miriam Katz, Jatja Fennel Mulligan, Benjamin S. Cramer, Marie Pierce Aubry, Robert A. Berner, Michael J. Novacek, and Warren M. Zapol. 2005. The rise of oxygen over the past 205 million years and the evolution of large placental mammals. *Science* 309: 2202–2204.
Francyk, Karen J., Janet K. Pitman, and Douglas J. Nichols. 1990. *Sedimentology, mineralogy, palynology and depositional history of some uppermost Cretaceous and lowermost Tertiary rocks along the Utah Book and Roan Cliffs east of the Green River.* U.S. Geological Survey Professional Paper 1787-N.
Fridell, Richard A., and Megan K. Morvilius. 2005. *Distribution and abundance of fish in the Virgin River between the Washington Fields Diversion and Pah Tempe, 2004.* Utah Division of Wildlife Resources, Publication Number 05-21.
Fridell, Richard A., Megan K. Morvilius, and Kevin K. Wheeler. 2004. *Inventory and distribution of fish in the Escalante River and tributaries, Grand Staircase–Escalante National Monument, Utah.* Utah Division of Wildlife Resources Publication Number 04-02.
Garvin, Christofer D., Thomas C. Hanks, Robert C. Finkel, and Arjun M. Heimsath. 2005. Episodic incision of the Colorado River in Glen Canyon Utah. *Earth Surface Processes and Landforms* 30, no. 8: 973–984.
Gilbert, C. R. 1993. Evolution and phylogeny. In *The physiography of fishes,* ed. David H. Evans, 1–45. Boca Raton, Fla.: CRC Press.
Gilbert, G. K. 1877. *Report on the geology of the Henry Mountains.* Geographical and Geological Survey of the Rocky Mountain Region (U.S.). Washington, D.C.: Government Printing Office.
Gilluly, James. 1929. *Geology and oil and gas prospects of part of the San Rafael Swell, Utah.* U.S. Geological Survey Bulletin 806.
Goetzmann, William H. 1966. *Exploration and empire.* New York: Alfred A. Knopf.
Goldstrand, Patrick M. 1994. *Tectonic development of Upper Cretaceous to Eocene strata of southwestern Utah.* Geological Society of America Bulletin 106.
Graf, William L. 1980. Fluvial processes in the lower Fremont River basin. In *Henry Mountains symposium,* ed. M. Dane Picard, 177–184. Utah Geological Association Publication no. 8.
Gualtieri, J. L. 1988. *Geologic map of the Westwater 30' x 60' quadrangle, Grand and Uintah counties, Utah, and Garfield and Mesa counties, Colorado.* U.S. Geological Survey Map I-1765.
Hansen, Wallace R. 1965. *Geology of the Flaming Gorge area, Utah-Colorado-Wyoming.* U.S. Geological Survey Professional Paper 490.
———. 1969. *The geologic story of the Uinta Mountains.* U.S. Geological Survey Bulletin 1291.
———. 1984. Post Laramide tectonic history of the eastern Uinta Mountains, Utah, Colorado, and Wyoming. *The Mountain Geologist* 21: 5–29.
———. 1985. Drainage development of the Green River basin in southwestern Wyo-

ming and its bearing on fish biogeography, neotectonics, and paleoclimates. *The Mountain Geologist* 22: 192–204.

———. 1986. *Neogene tectonics and geomorphology of the eastern Uinta Mountains in Utah, Colorado, and Wyoming.* U.S. Geological Survey Professional Paper 1356.

———. 1996. Dinosaur's restless rivers and craggy canyon walls. Vernal, Utah: Dinosaur Nature Association.

Harbor, David Jorgensen. 1998. Dynamic equilibrium between an active uplift and the Sevier River, Utah. *The Journal of Geology* 106: 181–194

Hayden, F. V. 1862. Some remarks in regards to the period of elevation of those ranges of the Rocky Mountains, near the sources of the Missouri River and its tributaries. *American Journal of Science*, second series, v. 33: 305–313.

Hayes, Phillip T. 1969. Yampa river supplement to *Powell Centennial*, Vol. 1. Denver: Powell Society.

Hayes, Phillip T., and Elmer S. Santos. 1969. *River runners' guide to the canyons of the Green and Colorado rivers with emphasis on geologic features. Powell Centennial, Vol. 1: From Flaming Gorge Dam through Dinosaur Canyon to Ouray.* Denver: Powell Society.

Hayes, Phillip T., and George C. Simmons. 1973. *River runners' guide to Dinosaur National Monument and vicinity, with emphasis on geologic features.* Denver: Powell Society.

Hepworth, Dale K., and Louis N. Berg. 1997. *Conservation management of native Bonneville cutthroat trout (Oncorhynchus clarki utah) in southern Utah.* Utah Department of Natural Resources Publication 97-05.

Hereford, Richard, Gordon C. Jacoby, and V. A. S. McCord. 1996. *Late Holocene alluvial geomorphology of the Virgin River in the Zion National Park Area, southwest Utah.* Geological Society of America Special Paper 310.

Hintze, Lehi F. 1988. Geologic history of Utah. Provo: Brigham Young University, Department of Geology.

Hochberg, Amy. 1996. Aminostratigraphy of Thatcher Basin, SE Idaho: Reassessment of Pleistocene Lakes. Masters thesis, Utah State University, Logan.

Howard, Alan D. 1986. *Quaternary landform evolution of the Dirty Devil River system, Utah.* Geological Society of America, Abstracts with Programs, v. 18, no. 6: 641.

Huber, Gary C. 1973. The canyon of the San Juan River. In *Geology of the canyons of the San Juan River*, ed. D. L. Baars, 8–15. Durango: Four Corners Geological Society.

Huber, P., S. G. Lucas, and A. P. Hunt. 1993. Late Triassic fish assemblages of the North American Western Interior. In *Aspects of Mesozoic geology and paleontology of the Colorado Plateau*, ed. M. Morales, 51–66. Museum of Northern Arizona Bulletin 59.

Hunt, Charles B. 1953. *Geology and geography of the Henry Mountains region, Utah.* U.S. Geological Survey Professional Paper 228.

———. 1956. *Cenozoic geology of the Colorado Plateau.* U.S. Geological Survey Professional Paper 279.

———. 1969. *Geologic history of the Colorado River.* U.S. Geological Survey Professional Paper 669-C.

———. 1987. Physiography of Western Utah. In *Cenozoic Geology of Western Utah*, ed. R. S. Kopp and R. E. Cohenour, 1–29. Utah Geological Association Publication 16.

Huntoon, Peter W. 1982. *Meander Anticline*. Geological Society of America Bulletin v. 93, p. 941–950.

———. 1988. Late Cenozoic gravity tectonic deformation related to the Paradox salts in the Canyonlands area of Utah. In *Salt deformation in the Paradox region*, 79–93. Utah Geological Survey Bulletin 122.

Huntoon, Peter W., George H. Billingsley Jr., and William J. Breed. 1982. *Geologic map of Canyonlands National Park and vicinity, Utah*. Moab, Utah: Canyonlands Natural History Association.

Irving, Washington. 1850. *The adventures of Captain Bonneville*. New York: Putnam.

Johnston, Robert E., and An Yin. 2001. Kinematics of the Uinta Fault system (southern Wyoming and northern Utah) during the Laramide Orogeny. *International Geology Review* 43: 52–68.

Kirkland, J. I. 1998. Morrison fishes. *Modern Geology* 22: 503–533.

Kowallis, Bart J., Eric H. Christiansen, Elizabeth Balls, Matthew T. Heizler, and Douglas Sprinkle. 2005. The Bishop Conglomerate ash beds, south flank of the Uinta Mountains, Utah: Are they pyroclastic fall beds from the Oligocene ignimbrites of western Utah and eastern Nevada? In *Uinta Mountain geology*, ed. Carol M. Dehler, Joel L. Pederson, Douglas A. Sprinkle, and Bart J. Kowallis, 131–145. Utah Geological Association Publication 33.

Kurlish, Richard A., III, and John J. Anderson. 1988. *The Sevier River formation (Miocene-Pliocene) of the Markagunt Plateau, Utah: Deposits unrelated to present physiography*. Geological Society of America, Abstracts with Programs, vol. 20, no. 5: 352.

Laabs, Benjamin C., and Eric C. Carson. 2005. Glacial geology of the southern Uinta Mountains. In *Uinta Mountain geology*, ed. Carol M. Dehler, Joel L. Pederson, Douglas A. Sprinkle, and Bart J. Kowallis, 235–253. Utah Geological Association Publication 33.

Laabs, Benjamin C., Jeremy D. Shakun, Jeffrey S. Munroe, David M. Mickelson, Bradley S. Sanger, Marc W. Caffee. 2003. *Cosmogenic exposure age limit on the last glacial maximum in south central Uinta Mountains, northeastern Utah*. Geological Society of America, Abstracts with Programs, vol. 35, no. 6: 541.

Lawrence, John C. 1963. Origin of the Wasatch Formation, Cumberland Gap area, Wyoming. *Rocky Mountain Geology* 2: 151–158.

Lawton, T. F., D. A. Sprinkle, P. G. DeCelles, P. G. Mitra, A. J. Sussman, and M. P. Weiss. 1997. Stratigraphy and structure of the Sevier thrust belt and proximal foreland-basin system in central Utah: A transect from the Sevier Desert to the Wasatch Plateau. In *Mesozoic to Recent Geology of Utah*, BYU Geology Studies, v. 42, part 2, p. 33–67, ed. Paul Karl Link and Bart J. Kowallis. Provo: Brigham Young University.

Lee, Willis T., Ralph W. Stone, Hoyte S. Gale, et al. 1916. *Guidebook of the western United States, part B: The overland route with a side trip to Yellowstone Park*. U.S. Geological Survey Bulletin 612.

Link, P. K., D. S. Kaufman, and G. D. Thackray. 1999. Field guide to Pleistocene Lakes Thatcher and Bonneville and the Bonneville flood, south eastern Idaho. In *Guidebook to the geology of eastern Idaho*, ed. S. S. Hughes and G. D. Thackray, 251–266. Pocatello: Idaho Museum of Natural History.

Lohman, S. W. 1974. *The geologic story of Canyonlands National Park*. U.S. Geological Survey Bulletin 1327.

References

Lyell, Sir Charles. 1853. *Principles of geology, ninth edition*. New York: D. Appleton.

Maisey, John G. 2000. Discovering fossil fishes. Boulder, Colo.: Westview Press.

Mann, Daven Craig. 1974. Clastic Laramide sediments of the Wasatch Hinterland, northeastern Utah. Masters thesis, University of Utah.

Marchetti, David William. 2002. Cosmogenic 3-helium exposure age dating of Pleistocene deposits in the Capitol Reef area, Utah. Masters thesis, University of Utah, Salt Lake City.

Marchetti, David W., and Thure E. Cerling. 2005. Cosmogenic ^{3}He exposure ages of Pleistocene debris flows and desert pavements in Capitol Reef National Park, Utah. *Geomorphology* 67: 423–435.

Marchetti, David William, Thure E. Cerling, and Richard B. Waitt. 2000. *Cosmogenic ^{3}He exposure age dating of a Pleistocene debris flow deposit, Capitol Reef National Park, Utah*. Geological Society of America, Abstracts with Programs, vol. 32, no. 7: 27.

Marchetti, David William, J. C. Dohrenwend, and T. E. Cerling. 2005. Geomorphology and rates of landscape change in the Fremont River drainage, northwestern Colorado Plateau. In *Interior Western United States: Geological Society of America Field Guide 6*, doi: 10.1130/2005.fld006(04), ed. J. Pederson and C. M. Dehler. Washington, D.C.: U.S. Geological Society.

Mattox, Stephen R., and Malcom P. Weiss. 1987. *Salt diapirism: An accessory to the structural development of Central Utah*. Geological Society of America, Abstracts with Programs, vol. 19, no. 5: 319.

McClane, A. J., ed. 1965. McClane's New Standard Fishing Encyclopedia. New York: Holt, Rinehart and Winston.

Mcfarland, William N., F. Harvey Pough, Tom J. Cade, and John B. Heiser. 1985. Vertebrate life. New York: Macmillan.

Milhous, Robert T. 2005. Climate change and changes in sediment transport capacity in the Colorado Plateau, USA. In *Sediment Budgets 2* International Association of Hydrological Sciences Publication 292, ed. Arthur J. Horowitz and Des Walling, 271–278.

Miller, David E. 1980. Great Salt Lake: A historical sketch. In *Great Salt Lake: A scientific, historical and economic overview*, ed. J. Wallace Gwynn, 1–14. Utah Geological and Mineral Survey Bulletin 116.

Miller, R. R. 1965. Quaternary freshwater fishes of North America. In *The quaternary of the United States: A review volume for the 7th congress of the International Association for Quaternary Research*, ed. H. E. Wright and D. G. Frey, 569–581. Princeton, N.J.: Princeton University Press.

Morgan, Dale L. 1947. *The Great Salt Lake*. Indianapolis: Bobbs-Merrill. Reprint, Salt Lake City: University of Utah Press.

———. 1953. *Jedediah Smith and the opening of the West*. Lincoln: University of Nebraska Press.

Morrill, Carrie, and Paul L. Koch. 2002. Elevation or alteration? Evaluation of isotopic constraints on paleoaltitudes surrounding Eocene Green River Basin. *Geology* 30: 151–154.

Morvilius, Megan K., and Richard A. Fridell. 2005. *Inventory and distribution of fish in the Escalante River and tributaries, Grand Staircase–Escalante National Monument, Utah 2003–2004*. Utah Division of Wildlife Resources Publication 05-15.

Munroe, Jeffrey S. 2005. Glacial geology of the northern Uinta Mountains. In *Uinta Mountain geology*, ed. Carol M. Dehler, Joel L. Pederson, Douglas A. Sprinkle, and Bart J. Kowallis, 215–234. Utah Geological Association Publication 33.

Mutschler, Felix E. 1969a. *River runners' guide to the canyons of the Green and Colorado rivers with emphasis on geologic features, Powell Centennial, Vol. 2: Labyrinth, Stillwater, and Cataract Canyons*. Denver: Powell Society.

———. 1969b. *River runners' guide to the canyons of the Green and Colorado rivers with emphasis on geologic features, Desolation and Gray Canyons*. Denver: Powell Society.

Nelson, Alan R., and Carol K. Krinsky. 1982. Late Cenozoic history of the upper Weber and Provo rivers, N.E. Utah. Geological Society of America, Abstracts with Programs, no. 6: 344.

Nichols, Gary C. 2002. *River runners' guide to Utah and adjacent areas*. Revised edition. Salt Lake City: University of Utah Press.

Norris, Richard D., Lawrence S. Jones, Richard M. Corfield, and Julie E. Cartlidge. 1996. Skiing the Eocene Uinta Mountains? Isotopic evidence in the Green River Formation for snow melt and large mountains. *Geology* 24: 403–406.

Oviatt, Charles G. 1987. *Probable late Cenozoic capture of the Sevier River into the Sevier Desert Basin, Utah*. Utah Geological Association Publication, v. 16: 265–269.

Patterson, C. 1981. The development of the North American fish fauna, a problem of historical biogeography. In *The evolving biosphere*, ed. Forey, P. L., 265–281. London: British Museum of Natural History and Cambridge University Press.

Patton, Peter C., and Paul J. Boison. 1986. *Processes and rates of formation of Holocene alluvial terraces in Harris Wash, Escalante River Basin, south-central Utah*. Geological Society of America Bulletin, vol. 97: 369–378.

Patton, Peter C., Norma Bigger, Christopher D. Dondit, Mary L. Gillam, David W. Love, Michael N. Machette, Larry Mayer, Roger B. Morrison, and John N. Rosholt. 1991. Quaternary geology of the Colorado Plateau. In *Quaternary nonglacial geology conterminous U.S.*, ed. Roger B Morrison, 373–406. Boulder: Geological Society of America.

Pederson, Joel L., and Kevin W. Hadder. 2005. Revisiting the classic conundrum of the Green River's integration through the Uinta uplift. In *Uinta Mountain geology*, ed. Carol M. Dehler, Joel L. Pederson, Douglas A. Sprinkle, and Bart J. Kowallis, 149–154. Utah Geological Association Publication 33.

Powell, Allan Kent, ed. 1994. Utah history encyclopedia. Salt Lake City: University of Utah Press.

Powell, John Wesley. 1875. *Report of J. W. Powell exploration of the Colorado River of the west and its tributaries explored in 1869, 1870, 1871, and 1872*. Washington D.C.: Government Printing Office.

———. 1876. *Report on the geology of the eastern portion of the Uinta Mountains and a region of country adjacent thereto*. Washington D.C.: U.S. Government Printing Office.

———. 1957. *The exploration of the Colorado River*. Abridged from the first edition of 1875. Chicago: University of Chicago Press.

Prettyman, Brett. 2001. Fishing Utah. Guilford, Conn.: The Globe Pequot Press.

Reheis, Marith C., Robert C. Palmquist, Sherry S. Agaard, Cheryl Jaworowski, Brainerd Mears, Jr., Richard F. Madole, Alan R. Nelson, and Gerald D. Osborn. 1991.

Quaternary history of some southern and central Rocky Mountain basins: Bighorn Basin, Green Mountain–Sweetwater River area, Laramie Basin, Yampa River basin, northwestern Uinta Basin (modified). In *Quaternary nonglacial geology: conterminous U.S.*, ed. Roger B. Morrison, 407–440. Boulder: Geological Society of America.

Repka, James L., Robert S. Anderson, and Robert C. Finkel. 1997. Cosmogenic dating of fluvial terraces, Fremont River, Utah. *Earth and Planetary Science Letters* 152: 59–73.

Riggs, N. R., T. M. Lehman, G. E. Gehreis, and W. R. Dickinson. 1996. Detrital zircon link between headwaters and terminus of the Upper Triassic Chinle-Dockum paleoriver system. *Science* 273: 97–100.

Rowley, Peter Dewitt. 1968. Geology of the southern Sevier Plateau, Utah. Ph. D diss., University of Texas, Austin.

Rowley, Peter D., Wayne L. Hamilton, William R. Lund, and David Sharrow. 2002. *Rockfall and landslide hazards of the canyons of the upper Virgin River basin near Rockville and Springdale, Utah.* Geological Society of America, Abstracts with Programs, vol. 34, no. 4: 50.

Rowley, Peter D., Thomas A. Steven, and Harald H. Mehnert. 1981. *Origin and structural implications of Upper Miocene rhyolites in Kingston Canyon, Piute County, Utah.* Geological Society of America Bulletin, Part I, 92: 590–602.

Ryden, Dale W. 2000. *Adult fish community monitoring on the San Juan River, 1991–1997.* Washington, D.C.: U.S. Fish and Wildlife Service.

Sears, Julian D. 1924. *Relations of the Browns Park Formation and the Bishop Conglomerate, and their role in the origin of the Green and Yampa rivers.* Bulletin of the Geological Society of America, vol. 35: 279–304.

Sharrow, David L. 2004. *Use of repeat photography and other techniques to understand river morphology and riparian vegetation changes along a channelized reach of the north fork of the Virgin River in Zion National Park.* Geological Society of America, Abstracts with Programs, vol. 36, no. 5: 125.

Smith, G. R. 1981. *Late Cenozoic freshwater fishes of North America.* Ann. Rev. Ecol. Supt. 163–193.

———. 1999. *Using fish paleontology to reconstruct drainage histories.* Geological Society of America, Abstracts with Programs, vol. 31, no. 7: 443.

Sorensen, F. Wendy, and C. Frederick Lohrengel II. 2001. *Sedimentation in paleolake Grafton, Zion National Park area, Utah.* Geological Society of America. Abstracts with Programs, vol. 33, no. 5: 59.

Spangler, Lawrence E. 2005. Geology and karst hydrology of the eastern Uinta Mountains: An overview. In *Uinta Mountain geology*, ed. Carol M. Dehler, Joel L. Pederson, Douglas A. Sprinkle, and Bart J. Kowallis, 205–214. Utah Geological Association Publication 33.

Sparks, B. W. 1986. *Geomorphology.* 3rd edition. New York: Longman.

Standlee, Larry A. 1982. Structure and stratigraphy of Jurassic rocks in central Utah: Their influence on tectonic development of the Cordilleran foreland thrust belt. In *Geologic Studies of the Cordilleran thrust belt*, ed. Richard Blake Powers. Denver, Colo.: Rocky Mountain Association of Geologists.

Steiner, Maureen Bernice. 2003. *Paleogeography of the Colorado Plateau region, Pennsylvanian through Cretaceous, and mid-Morrison plateau rotation.* Geological Society of America, Abstracts with Programs, vol. 35, no. 5: 34.

Steven, T. A., H. T. Morris, and P. D. Rowley. 1990. *Geologic map of the Richfield 1° ×
2° quadrangle, west-central Utah*. U.S. Geological Survey Map I-1901.

Stevens, John Dale. 1969. The High Uintas, a landform analysis. Ph.D. diss., University of California, Los Angeles.

Stevenson, Gene M. 2000. Geology of Goosenecks State Park, San Juan County, Utah.
In *Geology of Utah's Parks and Monuments*, ed. D. A. Sprinkel, T. C. Chidsey, and
P. B. Anderson, 433–447. Utah Geological Association Publication 28.

Stewart, J. D. 1996. Cretaceous acanthomorphs of North America. In *Mesozoic
fishes – systematics and paleoecology*, ed. G. Arratia and G. Viohl, 383–394. Proceedings of the international meeting, Eichstatt 1993, Verlag Dr. Friedrich Pfeil,
Munchen, Germany.

Stokes, W. L. 1978. *River capture in central Utah*. Geological Society of America, Abstracts with Programs, vol. 10, no. 5: 238–239.

Stokes, W. L., L. F. Hintze, and J. H. Madsen, compilers. 1961–1963. Geologic map of
Utah, scale 1:250,000. Utah Geological Survey.

Taylor, D. W., and R. C. Bright. 1987. Drainage history of the Bonneville Basin. In
*Cenozoic geology of western Utah: sites for precious metal and hydrocarbon
accumulation*, ed. R. S. Kopp and R. E. Cohenour, 239–256. Utah Geological Association Publication 16.

Turner-Peterson, Christine Elizabeth. 1987. Sedimentology of the Westwater Canyon
and Brushy Basin members, upper Jurassic Morrison Formation, Colorado Plateau, and relationship to Uranium mineralization. Ph.D. thesis, University of Colorado, Boulder.

Tyler, Noel, and Frank G. Ethridge. 1983. Depositional setting of the Salt Wash member of the Morrison Formation, southwest Colorado. *Journal of Sedimentary Petrology* 53: 67–82.

Tyus, H. M., B. C. Burdick, R. A. Valdez, C. M. Haynes, T. A. Lytle, and C. R. Berry.
1982. Fishes of the upper Colorado River Basin: Distribution, abundance, and
status. In *Fishes of the upper Colorado River system: Present and future*, ed. W. H.
Miller and H. M. Tyus, 12–70. Western Division, American Fisheries Society.

Urbanek, Mae. 1988. *Wyoming place names*. Missoula, Mont.: Mountain Press.

Utah Department of Natural Resources. 1991. *Hydrologic inventory of the Sevier
River Basin*. Utah Department of Natural Resources, Division of Water Resources.

Van Cott, John W. 1990. *Utah place names*. Salt Lake City: University of Utah Press.

Villien, A., and R. M. Kligfield. 1986. Thrusting and synorogenic sedimentation in
central Utah. In *Paleotectonics and sedimentation in the Rocky Mountain Region*, ed. James A. Peterson, 281–307. AAPG Memoir vol. 41.

Wachtell, Douglas Lowell. 1988. Sedimentology and tectonic setting of conglomerate
in Paleocene North Horn Formation, Spanish Fork Canyon and Mount Nebo areas, Central Utah. Masters thesis, University of Utah.

Waitt, Richard B. 1997. *Huge Pleistocene-(Pliocene?) debris avalanches from Aquarius Plateau through Waterpocket Monocline, Utah*. Geological Society of America, Abstracts with Programs, vol. 29, no. 6: 410.

Wallace, Chester Alan. 1972. A basin analysis of the Upper Precambrian Uinta Mountain Group, Utah. Ph.D. diss., University of California, Santa Barbara.

Warner, Ted J., ed. 1976. The Domínguez–Escalante journal. Trans. Fray Angelico
Chavez. Provo: Brigham Young University Press.

Webber, C. E., and C. M. Bailey. 2003. *Structural and fluvial history of the Fish Lake basin, High Plateaus, central Utah*. Geological Society of America, Abstracts with Programs, vol. 35, no. 4: 69.

Wilford, John Noble. 2000. *The mapmakers*. New York: Alfred A. Knopf.

Williams, Norman C., and James H. Madsen Jr. 1959. Late Cretaceous stratigraphy of the Coalville area, Utah. In *Intermountain Association of Petroleum Geologists Guidebook,* 10th annual Field Conference.

Willis, Grant C. 1986. *Geologic map of the Salina quadrangle, Sevier County, Utah*. Utah Geological Survey map 83.

———. 1991. *Geologic map of the Redmond Canyon quadrangle, Sanpete and Sevier counties, Utah*. Utah Geological Survey map 138.

———. 2000. *Knowledge of Utah thrust system pushed forward*. Utah Geological Survey, Survey Notes, vol. 29, no. 1: 1–4.

Wilson, Mark V. H., and John D. Bruner. 2004. Mesozoic fish assemblages of North America. In *Mesozoic Fishes, Vol. 3, Systematics, Paleoenvironments and Biodiversity*, ed. Gloria Arratia and Andrea Tintori, 575–595. Proceedings of the International Meeting, Serpiano, Verlag, Dr. Frederich Pfeil, Munchen, Germany.

Winkler, Gary Ralphs. 1970. Sedimentology and geomorphic significance of the Bishop Conglomerate and the Browns Park Formation, Eastern Uinta Mountains, Utah, Colorado, and Wyoming. Masters thesis, University of Utah.

Witkind, Irving J. 1992. Paired, facing monoclines in Sanpete–Sevier Valley area, Central Utah. *The Mountain Geologist* 29: 5–17.

Witkind, Irving J., and William R. Page. 1984. Origin and significance of the Wasatch and Valley Mountains monoclines, Sanpete–Sevier Valley area, Central Utah. *The Mountain Geologist* 21: 143–156.

Witkind, Irving J., and Malcolm P. Weiss. 1991. *Geologic map of the Nephi 30' × 60' quadrangle, Carbon, Emery, Juab, Sanpete, Utah, and Wasatch counties, Utah*. Miscellaneous Investigations Series, U.S. Geological Survey Report I-1937.

Wolkowinsky, Amy J., and Darryl W. Granger. 2004. Early Pleistocene incision of the San Juan River, Utah, dated with ^{26}Al and ^{10}Be. *Geology* 32: 749–752.

Index

Page numbers printed in italics and followed by
f, m, or t refer, respectively, to figures, maps, or tables.

INDEX

214